Routledge Introductions to Environment

Gender and Environment

Susan Buckingham-Hatfield

London and New York

First published 2000 by Routledge
2 Park Square, Milton Park, Abingdon, Oxon, OX14 4RN

Simultaneously published in the USA and Canada
by Routledge
270 Madison Ave, New York NY 10016

Routledge is an imprint of the Taylor & Francis Group

Transferred to Digital Printing 2006

Typeset in Times by Keystroke, Jacaranda Lodge, Wolverhampton

British Library Cataloguing in Publication Data
A catalogue record for this book is available from the British Library

Library of Congress Cataloging in Publication Data
Buckingham-Hatfield, Susan.
    Gender and environment / Susan Buckingham-Hatfield
        p.   cm.
    Includes bibliographical references (p.).
    1. Sex role–Environmental aspects.   2. Man-woman relationships–
    Environmental aspects.   I. Title

HQ1075.B8 2000
305.3–dc21   99–054628

ISBN 0–415–16819–8 (hbk)
ISBN 0–415–16820–1 (pbk)
Printed and bound by CPI Antony Rowe, Eastbourne

Contents

Series editors' preface: *Environment and Society* titles

The 1970s and early 1980s constituted a period of intense academic and popular interest in processes of environmental degradation: global, regional and local. However, it soon became increasingly clear that reversing such degradation would not be a purely technical and managerial matter. All the technical knowledge in the world does not necessarily lead societies to change environmentally damaging behaviour. Hence a critical understanding of socio-economic, political and cultural processes and structures has become, it is acknowledged, of central importance in approaching environmental problems. Over the past two decades in particular there has been a mushrooming of research and scholarship on the relationships between social sciences and humanities on the one hand and processes of environmental change on the other. This has lately been reflected in a proliferation of associated courses at undergraduate level.

At the same time, changes in higher education in Europe, which match earlier changes in America, Australasia and elsewhere, mean that an increasing number of such courses are being taught and studied within a framework offering maximum flexibility in the typical undergraduate programme: 'modular' courses or their equivalent.

The volumes in this series will mirror these changes. They will provide short, topic-centred texts on environmentally relevant areas, mainly within social sciences and humanities. They will reflect the fact that students will approach their subject matter from a great variety of different disciplinary backgrounds; not just within social sciences and humanities, but from physical and natural sciences too. And those students may not be familiar with the background to the topic, they may or may not be going on to develop their interest in it, and they cannot automatically be thought of as being at 'first-year level', or second- or third-year: they might need to study the topic in any year of their course.

The authors and editors of this series are mainly established teachers in higher education. Finding that more traditional integrated environmental studies or specialised academic texts do not meet their requirements, they have increasingly met the new challenges caused by structural changes in education by writing their own course materials for their own students. These volumes represent, in modified form which all students can now share, the fruits of their labours.

To achieve the right mix of flexibility, depth and breadth, the volumes, like most modular courses themselves, are designed carefully to create maximum accessibility to readers from a variety of backgrounds. Each leads into its topic by giving adequate introduction, and each 'leads out' by pointing towards complexities and areas for further development and study. Indeed, much of the integrity and distinctiveness of the Environment and Society titles in the series will come through adopting a characteristic, though not inflexible, structure to the volumes. Each introduces the student to the real-world context of the topic, and the basic concepts and controversies in social science/humanities which are most relevant. The core of each volume explores the main issues. Data, case studies, overview diagrams, summary charts and self-check questions and exercises are some of the pedagogic devices that will be found. The last part of each volume will normally show how the themes and issues presented may become more complicated, presenting cognate issues and concepts needing to be explored to gain deeper understanding. Annotated reading lists are important here.

We hope that these concise volumes will provide sufficient depth to maintain the interest of students with relevant backgrounds, and also sketch basic concepts and map out the ground in a stimulating way for students to whom the whole area is new.

The Environment and Society titles in the series complement the Environmental Science titles which deal with natural science-based topics. Together this comprehensive range of volumes which make up the Routledge Introductions to Environment Series will provide modular and other students with an unparalleled range of perspectives on environmental issues, cross referencing where appropriate.

The main target readership is introductory level undergraduate students predominantly taking programmes of environmental modules. But we hope that the whole audience will be much wider, perhaps including second- and third-year undergraduates from many disciplines within the social sciences, science/technology and humanities, who might be taking occasional environmental courses. We also hope that sixth-form teachers and the wider public will use these volumes when they feel the need to obtain quick introductory coverage of the subject we present.

David Pepper and Phil O'Keefe
1997

Series International Advisory Board

Australasia: Dr P. Curson and Dr P. Mitchell, Macquarie University

North America: Professor L. Lewis, Clark University; Professor L. Rubinoff, Trent University

Europe: Professor P. Glasbergen, University of Utrecht; Professor van Dam-Mieras, Open University, The Netherlands

Figures

Tables

Boxes

Author's preface

Feminist research holds that we should all acknowledge our own position in the research and writing that we do, so that we, and our readership, can be aware of our potential biases and can get an idea of where we are coming from.

I have been teaching environmental issues, political and urban geography in British universities for around fifteen years. I began teaching in the year in which the Women and Geography Study Group of the Institute of British Geographers (now merged with the Royal Geographical Society) published their landmark book *Geography and Gender*, by which I was profoundly influenced. Increasingly I have incorporated gender concerns and feminist perspectives into my teaching, research and writing. The reaction to this from my students is mixed: some, particularly the younger students for whom university is their first break from the parental home, are sceptical that there still exist gender divisions. They have done well in A levels, enjoy an egalitarian relationship with their male and female peers, and have few responsibilities to weigh them down. Older students, some of whom have children, who have to balance many activities, and who may not have sailed through school, give example after example of how the gender divide persists and quickly grasp the implications of this for society environment relations.

I do not have children and I have a secure and reasonably well-paid job in a university. I do not know how my (particularly female) colleagues cope with the exigencies of a demanding career, domestic and childcare responsibilities. My class (middle), age (also middle) and race (white) insulate me from many of the scenarios I present in this book and yet I am constantly amazed and upset by the experiences which reach me secondhand from friends, colleagues and family which support the persistence of unequal gender relations and the environmental incivilities this exposes women to. I will give two examples: the first is contemporary and was relayed to me by a colleague last week whose girlfriend had been profoundly distressed by the birth of her third daughter when her husband and her family had vociferously wanted a boy child. The second is more historical and concerns my maternal grandmother who, I recently learned (I unfortunately never met her), was the eldest of twelve children and had to collect water from a local river and carry it back to her home in buckets hanging from a yolk across her shoulders. We tend to think of events and lifestyles in the 'developing' world as very far from ours, and yet these two anecdotes show this to be far from the truth.

I began this book on a sabbatical from teaching in 1997 and was promptly forced to abandon the physical process of writing it by a horseriding accident which put me in hospital for a month. In fact this gave me invaluable time to think the project through more carefully, to miss and value the company of my colleagues and students, and to be an observer of another set of gender inequalities in the medical profession (in which the hard-worked and underpaid, exclusively women, nurses had an equally responsible job as the hard-worked, but fairly paid and exclusively men doctors).

I would like to dedicate this book to those nurses and doctors, to all my students – men and women – who continue to be a great pleasure and inspiration to me, and to young people everywhere in whose hands the future lies.

Susan Buckingham-Hatfield,
London, 1999

Acknowledgements

In addition to the students who attended my seminars on gender and who helped me sharpen my arguments, I would specifically like to thank my Head of Department, Callum Firth, who gave me teaching relief to write this book and who tries to run a department based on firm principles of equal opportunities, to all my other colleagues in the Department, but especially Simon Batterbury for discussions on gender and development, and to Fiona Smith, Kate Theobald and Iris Turner for discussions and collaborative research into gender and opportunities for post-graduate students. I would also like to thank Professor John Sumpter at Brunel for his help in understanding oestrogen pollution, in which he has an international reputation. I have participated to varying degrees in a number of networks and have got far more out of these than I have put in and would like to thank the ESRC Mainstreaming Gender in Public Policy Network run from Sheffield Hallam University, Eurofem (from which the ESRC network evolved), the Planning and Environment Research Group, the Women and Geography Study Group, and the Women's Environmental Network. Comments from reviewers of an early version of the manuscript were extremely helpful and I would like to extend a particular thank you to one of the Series Editors, David Pepper, for his enthusiasm and constructive comments at various stages of preparation. In addition I would like to thank my friends and family who were generous with moral support, particularly Geraldine Lievesley for discussions on women and politics and plans for future projects, and Richard Kevern for donating the photograph of young women carrying fuel in India. Other material has been reproduced from the Women's Environmental Network, The Ontario Advisory Council on Women's Issues, the Royal Botanic Gardens at Kew, the *Observer* and the *Independent on Sunday*, to all of whom I extend my thanks for permission to reproduce their material. Whilst I have drawn heavily on all these people for support, I take all responsibility for views portrayed in this book and for all its shortcomings. Finally I would like to acknowledge the inspiration and support I have received from Judith Matthews in developing this and other projects, and, although she will not read this (she died last year) I am sure she knows it.

1 ▶ Linking gender and the environment

Over the past ten years or so, the relationship between gender and the environment has become more explicit and apparent. The literature on environment and development has particularly stressed that women's work is often linked to the environment (through subsistence agriculture, domestic chores and hired work such as sowing and weeding) and that much of this work is made harder through environmental degradation. For example, as forests are decimated and sources of groundwater are depleted, women have to make longer, more time-consuming journeys to collect water and firewood. Most textbooks on environment and development will now refer to the discontinuities between the amount of work women do on the land and their lack of ownership of that land, a disparity which can reach almost feudal proportions.

The relationship between gender and the environment is less obvious in the West where most people are more distant from the source of their food supply, the energy and the water they use. However, because of their biology, it is women who conceive, carry, give birth to and suckle children and this exposes them to a number of environmental hazards. In addition, their social role as the main unpaid domestic worker in each household brings them closer to an awareness of environmental hazard, whether it is by shopping for food (women need to be aware of whether this food has been sprayed by harmful pesticides, irradiated for preservation or genetically modified), preparing that food for safe eating, or caring for the health of their children.

The growing literature on environmental justice argues that those people and communities who are most poor, or most undervalued, are most likely to be exposed to environmental hazard (see, for example, Bullard, 1999, and Dobson, 1999). For example, toxic industries tend to locate in communities which have the least clout in opposing this, whether this is a poor neighbourhood in a rich country, or in the developing world, with its weaker health and safety regulations and hunger for jobs and foreign exchange. Whilst it is usually the wealthy who pollute most perniciously (both individually and nationally), it is the poor who find it harder to escape from the negative effects of pollution. This literature has mostly focused on ethnic minorities and communities in poverty, since these groups have a clear spatial manifestation. Whilst women are fairly homogeneously distributed through the community, it is striking how, world-wide, women are more likely to be poor than men. Particularly where they are lone mothers or elderly, women are more likely to form a greater proportion of a community in poverty than are men.

Despite many differences between women in the North and South, there persist global inequalities in income and occupation between women and men. It is noticeable how women world-wide are engaged in those jobs which, whilst the most critical to human survival (feeding, clothing and providing health care), are the least valued in society. Since poverty is a major determinant of ill health and of exposure to many environmental problems, women are more likely to suffer these. That such income and occupation imbalances seem so resilient to geographical location and to industrial progress implies that society must be structured in such a way as to perpetuate these inequalities.

I would like to open this book with some comments on how our society has created and perpetuated this inequality. The system of industrial capitalism which has existed in the West for the past two hundred years, and which the West has exported to most parts of the world in the twentieth century, is built upon a set of social relations in which power and influence is associated with capital.

Although working-class women have undertaken paid work throughout this time, it was seen as unfitting for middle- and upper-class women to do so. Their responsibility was to manage the household and, in some cases, to extend this responsibility into the local community (by providing for the local poor, for example). Women had no role in the public arena of politics and the economy and were allowed no say in how civil society was arranged. Women were not permitted to vote until 1918 in the UK, 1920 in the USA and as recently as 1972 in Swiss national elections and 1986 in Liechtenstein (Randall, 1987), nor were they admitted to public institutions such as the Royal Academy (see Chapter 2). Although there were a number of universities which admitted women to women-only colleges, Oxford and Cambridge did not permit woman to graduate with a degree until the 1940s.

Working and studying in a university, it is easy to forget how short women's tenure has been in higher education, and also how unrepresentative we are of the population as a whole, although Chapter 2 will show how women lack career advancement in higher education, compared to men. I am certainly conscious of how out of touch I am with many of the women I write about: I have a reasonably well-paid job (salary on a par with my male colleagues of equivalent position) with a relatively secure pension. In discussions on gender inequality, it is not uncommon for students to challenge my suggestion that women are less privileged than men, experiencing as they do communal, co-ed, domestic arrangements and chores. They are optimistic that their generation will be different and that women and men equally will be able to choose their lifestyles, incomes and consumption patterns. It does not, however, take much imagination to look beyond the daily realities of their own lives to see how gender inequalities persist in the lives of their families, their friends who have not been to university, or in those communities to which they travel in their summer vacation in search of the exotic bargain holiday. I am also depressingly reminded of how this optimism can quickly fade when I catch up with women graduates who have become parents within a few years of graduation (especially those who are lone parents). Why is it, then, that even in our current Western, post-industrial society, where the proportion of women in higher education is now overtaking men, that there are still gender inequalities? Perhaps a useful way to open this discussion is to look at the nature of gender itself.

Defining gender

The simplest explanation of gender is 'that it is a social construction organised around biological sex. Individuals are born male or female, but they acquire over time a gender identity, that is what it means to be male or female' (Gregson et al., 1997: 53).[1] This meaning implies two different kinds of relationships: that between the two genders and that between gender and society. Because gender is a society's interpretation of maleness and femaleness, that society will determine what should be male and female characteristics and roles. Girls and boys growing up in that society are, therefore, encouraged to adopt these characteristics and to fulfil these roles. They will be rewarded for being appropriately feminine or masculine and this in turn helps to reinforce their behaviour. Assuming, for the moment, that there is a biological state that is immutably male or female (but, as Chapter 4 will show, this is challenged by some writers), gender is that package of expectations which a society associates with each sex. In Western society, which is predominantly Judeo-Christian, characteristics which are valued as male are: assertiveness, dominance, competitiveness, aggression and logic. These kinds of behaviour are nurtured in play and in study, just as compassion, cooperation and emotion are rewarded in girls. So, whilst we all may be born into a particular sex, we acquire and are socialised into a set of behaviours and characteristics which constitute gender. Because this socialisation informs everything that we do and experience, such as through school, the media and community organisations, it has pervasive long-term effects which cross generations.

It is apparent that these qualities, characteristics and forms of behaviour are not isolated, but defined in relationship to each other. That is, what is female is specifically that which is not male, and vice versa. (Chapter 2 will look at the emergence and endurance of these dualisms.) It is mean-ingless, therefore, to talk about one gender without reference to the other. Also, because gender is socially constituted, in talking about gender, we are talking about the relationship of women and men to society.

The remainder of this chapter will consider how the gender roles prevalent in Western society have developed over time and how the concept of patriarchy may be used to explain their persistence. I will then look at how the way a society is organised has implications for its relationship with the environment, which will demonstrate how gender and environment become linked. The final section in the chapter will explain the organisation of this book.

Development of gender roles

Chapter 2 looks in more depth at the way in which Western society (from the Greeks onward) has considered men to be superior to women because of the former's supposed ability to distance himself from nature. Enlightenment thought, which was a crucial intellectual foundation for industrialisation, believed that it was both possible and desirable not only to distance society from nature, but to use this distance to control nature. At the same time, there was a spatial dimension to the division between men and women, so that women inhabited the private sphere of the household, whilst men had freedom throughout the public sphere (these concepts will be developed in Chapters 5 and 6), from which women, particularly middle- and upper-class women, were excluded.

As industrialisation developed through the nineteenth century, working-class women were drawn into waged work, although their range of opportunity was limited. This work was frequently typified by the tasks women were accustomed to performing at home, and was characterised by low wages, which were often below subsistence levels (Bradley, 1989). Bradley argues that the 'characteristic features of women and men's work, as we know them today, were becoming the norm in 1861' (ibid.: 344).

The gendered division of labour in paid work takes place both horizontally (where different occupational sectors are associated with a particular sex) and vertically (where certain positions across sectors are held by one sex). For example, in the UK, in 1981, 95 per cent of all jobs held in the mining and quarrying sector were held by men (horizontal segregation), whilst 90 per cent of all secretarial, typing, receptionist, cashier, nursing, maid and canteen assistant jobs were held by women (vertical segregation) (Bradley, 1989).

There are various explanations as to how and why this job and associated income segregation exists and persists. Marxist analysis argues that women form what is termed a 'reserve army of labour' which is brought into the waged labour force at times of labour shortage, only to be 'let go' when there is a labour surplus. An example of this can be found during the two world wars during which women were drafted into the munitions factories and essential services to replace the men who had been conscripted into the armed forces. More recently, in the late 1980s, a direct appeal was made to women, particularly older women, to consider returning to public-sector work, whilst just a few years later some observers were advocating a return of women to the household to allow men to take the jobs available. This is a structural argument which looks at the way economic forces structure people's decisions and 'choices' in a predictable pattern.

Other explanations are more particular and are linked to lifestyle (women take jobs to fit in with their domestic arrangements); to the specific requirements and prejudices of employers; to social and human capital (where job opportunities depend on people's access to networks, which, in turn offer access to employers and information); and spatial constraints which make it more difficult for women, who are poorer and less likely to drive or have access to a car, to travel (Hanson and Pratt, 1995).

These explanations invite the questions: why are women constrained by household arrangements? Why are employers prejudiced against women workers? Why are men more able to use networks and why are men more mobile? Why, indeed, are secretaries paid less than miners, or part-time workers paid a lower rate than full-time workers, and why do fewer women belong to a trade union (through which wage rises are negotiated) than men? Another structural argument which attempts to grapple with these conundrums is one based on patriarchy, which Walby defines as

> a system of interrelated social structures through which men exploit women . . . The key sets of patriarchal relations are to be found in domestic work, paid work, the state and male violence and sexuality, while other practices in civil society have a limited significance.
>
> (Walby, 1986: 50–51)

Indeed, Walby and Bagguley see patriarchy and capitalism as two separate but interacting systems (in Hanson and Pratt, 1995: 12). Linda McDowell also argues that patriarchy and class need to be seen as mutually reinforcing as it is in the interests of capital to minimise wage costs whilst ensuring that it does not bear the costs of social and generational reproduction, that is the bearing and rearing of children, and the provision of daily needs such as food, clothing, health and warmth. (See McDowell, 1986 for a fuller explanation of why she believes it is in the interest of capital to pay and value women's work at a lower rate than men's.)

The interlinkage between employment or economic and social factors is stressed by Hanson and Pratt who explain how the job choices that people make are made primarily within segments of the job market – that is, that the gendered structure of employment constrains people's choices. This will help to perpetuate the remarkably consistent pattern of employment for men and women since the early nineteenth century (Bradley, 1989).

It is important to point out here that employment opportunities and constraints differ between women, as well as between women and men. For example, the gains in employment rights, equal pay and equal opportunities have largely benefited middle-class, professional women, whilst working-class women, often working in less well-regulated employment sectors, have seen little improvement. Also, whilst 17 per cent of white women in paid employment work in managerial or professional jobs in the UK, only 2 per cent of Afro-Caribbean women do. Moreover, the work that black women are engaged in is likely to be in the lowest-status jobs in nursing, hospital domestic work and hotel and catering work. These are likely to be back-room jobs where the women are least visible (Bradley, 1989).

Society, gender and environment

This form of industrial capitalism, which is deeply ingrained not only in the West, but, through the activities of transnational companies, globally, has a particular relationship with the environment, which is distinct from pre-industrial societies. As Chapter 2 will show, this is a relationship in which society considers nature to be a resource to be extracted and a waste dump limited only by our technological ingenuity (Simon, 1994).

Because society is structured in such a way that people are rewarded financially according to the value society accords their occupation and because access to amenity resources such as clean air, or peace and quiet, is governed by one's ability to pay, relative poverty will result in a lack of access to these resources. It will also, generally result in poorer people being exposed to proportionately higher levels of pollution and other negative environmental effects. Whilst considerable material gains are made through the exploitation of nature (such as energy, heating, continuously available food, better health care and provision), for each benefit, there is likely to be a negative externality or disbenefit so that, for example, one person's mobility is another's ill health linked to car pollution; for year-round availability of exotic foods, others suffer through the loss or degradation of farmland for subsistence; and for a plentiful supply of electricity, people suffer the negative effects of radioactive pollution or emphysema. The gains, then, are not evenly experienced by all members of society, and neither are the problems.

There are many different groups who will experience these problems in greater measure, for example, all those on low incomes, people and communities from ethnic minorities and the disabled (see the environmental justice literature mentioned earlier). Moreover, on a global scale, the South bears a significant environmental burden in order that the

West may reap the benefits (Guha and Martinez-Alier, 1997). Whilst this book acknowledges this and will refer to some of the more glaring inequalities, it is primarily concerned with gender, and will consequently be mainly looking at how these externalities are experienced differently between men and women. 'Women', of course, is a very broad category of people, and a woman's social, economic and environmental situation will differ depending on a number of factors: ethnicity, class, income, education, able-bodiedment, age, sexuality, nationality and motherhood. Although it is sometimes difficult to generalise about women, there are, nevertheless, powerful indicators that women do share a number of experiences which make some generalisation valid. First, they are the only sex that can bear children (despite some of the more lurid accounts from the *News of the World* and *National Enquirer*!), although a rising proportion of women are choosing not to do so. Secondly, there are very marked global figures on the disparities in income and occupation, and on political representation, which cross national boundaries. Moreover, household tasks are also remarkably similar between cultures, even though the equipment used to complete these, and the household forms may differ. Consequently, because of these experiences, women tend to experience environmental problems differently from men. At the same time, they are less likely to be in a position to affect these issues. Much environmental damage is a result of decisions taken by companies and governments and amelioration would also need to be taken by them. Women form a very small percentage of decision makers in these organisations.

In this introduction to gender and environment, I make no apology for focusing on the similarities women encounter, but I would urge the reader to contemplate how women of different educational backgrounds, incomes, ethnicities and so on might experience environmental problems differently and I point readers to literature which develops this.

Structure of the book

The following two chapters introduce the way in which research into the environment has been influenced by gender relations, and reviews work which examines the problems with this, and possible solutions. Chapter 2 looks at the history of the way in which science has been practised, particularly over the past 300 years, in the West, and explores how this has been shaped by the way in which society has been organised. I look at key disciplines in the study of the environment to see how both the subject matter and methodologies used affect the practice of science (as well as the milieu in which it practised). This chapter concludes with some alternative methods for the practice of scientific inquiry which can help to avoid the gender bias currently inherent in most research.

Chapter 3 investigates how a recognition of gender roles, and a subsequent revaluing of these, might have a positive impact on the relationship between society and its environment. I focus here on a body of research which has emerged in the last twenty years, known as eco-feminism, which attempts to bring together the concerns of ecology and feminism. I review the two main strands of eco-feminist thought (cultural and social) and conclude the chapter with some critiques of these philosophies.

Chapters 4 to 7 consider specific environmental problems in two ways: (1) through the differential impact on men and women, and (2) because women and men have different capacities to affect decisions which impact on the environment. Chapter 4 focuses on the way in which male and female bodies experience environmental problems differently. Some of the reasons for this are linked to the biological make-up of the body, but others are due to the uses to which we put our bodies as a result of the roles we are socialised into playing by society. Chapter 5 concentrates on these social roles and focuses on gender roles within the family and/or household. Here, I consider how environmental impacts on women and men differ, depending on the roles they play. In Chapter 6, I begin to look in more depth at the structures in which decisions about the environment are made and at how these are influenced by gender relations. Since the more influential decisions concerning the environment are taken by bodies in the public arena of industry and politics, the interests governing these decisions are likely to reflect the class, gender and race of the decision makers. I look at how current structures of decision-making in the planning profession and politics favour men and masculinity. I also show the limited impact that women have in environmental action by examining how women's active presence at the grassroots of environmental protest is eclipsed by their notable lack of presence in senior staff appointments in environmental organisations. This is a reflection of employment and income patterns elsewhere in society, which is also reviewed here.

In Chapter 7, I look at the global dimension of gender–environment relations. For example, the global gender division of labour is a mirror of local and household patterns and I explore the impact that this has on the health of women and environments. International agencies, such as the United Nations, are major players in this global relationship (though not nearly as powerful as global capital), and I consider their ability, and inclination, to address gender inequalities in the light of environmental problems. Finally, the concluding chapter will attempt to thread these scales and themes together to explore how the way in which we interpret gender–environment relations determines the way in which we explain environmental problems and attempt to resolve them.

Although this book focuses on a number of issues which unite women's experience, such as lower incomes, child-bearing and rearing and domestic roles, I have tried, throughout the book, to show how women in different positions may experience environmental problems differently, particularly black women and women in poverty. In order to explore these differences, and other issues addressed in each chapter, more deeply, I refer the reader to a number of key texts at the end of each chapter. Some of these references may also be used to read more about other systems of domination in society (especially by race and by colonisation) which affect the relationship between different groups of people, and between these groups and the environment.

Discussion questions

1 Thinking of your family home, how are tasks distributed according to gender? Is this different in the new household you may have established whilst at university or college?

2 How has your upbringing (both within your family and, more broadly, in school and the community) emphasised your masculinity or femininity?

3 Consider how differences in educational background, income, ethnicity and ablebodiement cause women to experience environmental problems differently.

Note

1 A social construction is the way in which a particular society constructs, or creates, meaning.

2 ▶ The making of science: it's a man's world

Key words: dualisms; division of academic labour; fieldwork; gendering of science; objectivity/subjectivity; feminist methodologies

- Origins of Western scientific thought
- Female and male roles in the practice of science
- The gendering of science
- Redressing the gender bias in science

There is no one way to understand the world. That is, rather than there being a set of 'truths' waiting to be uncovered, our environmental curiosity is driven by prevailing social, political and moral frameworks. A society which is stratified by economic wealth, race or gender will seek to understand the environment in a way that reflects this. The questions asked, the methods used in seeking answers, and the answers themselves, are likely to reinforce and justify the way in which that society is organised.

In this chapter I will focus on the Western tradition of social structures and their interaction with nature. Not because it is the only, or the best, tradition but because it is, now, on the cusp of the twenty-first century, so globally pervasive, reaching into the most remote communities. The roots of this tradition are commonly traced back to Greek philosophy, although the Enlightenment, and particularly the work of Francis Bacon, is seen as forming the basis of modern Western science (Pepper, 1996: 143). After brief summaries of these foundations, this chapter will demonstrate how the practice and content of science has been affected by the fact that it has been predominantly practised by men and how science is consequently gendered in its methods, theories and outcomes. Many commentators, particularly those from a feminist tradition (such as Sandra Harding, Donna Haraway and Evelyn Fox Keller), now argue that this produces 'bad' science and the final part of the chapter will focus on methodologies which are being developed to challenge what is seen as gendered knowledge, to produce a better, or more honest science.

Origins of Western scientific thought

The tradition of scientific thought in the West owes its origins to Greek philosophy. As this philosophy developed, 'man's' ability to reason was seen as a way of distancing 'himself' from nature. This distancing became more apparent with the development of the 'dualisms' which originated in Pythagorean thought. Genevieve Lloyd writes of this evolution, from the early Greeks who saw a connection between women's capacity to conceive with the fertility of nature, to the Pythagorean relegation of femaleness to inferiority (1993: 42). As Box 2.1 shows, certain qualities were considered to be superior to others, with the female listed with other less desirable qualities (shown on the right-hand side of the table) associated with vagueness and formlessness. This is the origin of the dualisms – or opposites – which will be developed in Chapter 3 in relation to eco-feminism.

The qualities on the left-hand side are associated with clarity and form, which Pythagorean thought valued. The qualities on the right-hand side are manifestations of irregularity and disorder with which females were associated (Lloyd, 1993: 42).

Implicitly, qualities itemised in the left-hand column of Box 2.1 were considered superior to those in the right, and in so doing 'femaleness' was confirmed as an inferior quality to maleness. Such a conceptualisation originated from a society which valued order, a society which was structured in a particularly gendered way, where women were confined to

---

Box 2.1

Pythagorean opposites

Limit	Unlimited
Odd	Even
One	Many
Right	Left
Male	Female
Rest	Motion
Straight	Curved
Light	Dark
Good	Bad
Square	Oblong

the private sphere with no rights in the public or civic domain of the much-vaunted 'democratic' society (which, incidentally, also kept slaves).

Both Plato's introduction of the idea of objectivity, which distinguished the 'knower' from what might be 'known' and Aristotle's claim that men's superior intellect resulted from their superior physical make-up (with the corollary that women, whose bodies were considered inferior, therefore had inferior minds) have been powerful influences in the subsequent gendering of the practice of science and production of what was considered to be valued knowledge (which may exclude non-scientific knowledge, a point which will be developed in subsequent chapters).

Western philosophy has inherited these pervasive ideas, of dualistic thought and the potential for objectivity, from Greek philosophy and both notions informed Francis Bacon, generally considered the 'father' of modern scientific thought. However, Bacon rejected the idea that anything was unknowable. His method of empirical observation was rigorous; he warned against generalising from a limited number of cases and against manipulating 'facts' to support such generalisations (Bacon, 1620). He was also suspicious of the scientist researching into areas they attached personal importance to, as he (sic) may be in danger of biasing his results to suit his preference.

> let every student of nature take this as a rule – that whatever his mind
> seizes and dwells upon with peculiar satisfaction is to be held in
> suspicion, and that so much the more care is to be taken in dealing with
> such questions to keep the understanding even and clear.
>
> (ibid.: 1637)

With these precepts, then, the ground was prepared for a supposedly value-free science, the problems with which will be discussed later in the chapter.

There was also a purpose to acquiring knowledge through empirical observation: that was to dominate and to control. 'Nature' was still associated with the female, but these two concepts – nature as female and nature that was to be dominated – culminated in a graphic image of ravishment, although Bacon himself argued that nature could best be utilised if society worked in harmony with 'her' laws. Chapter 3 will develop the links between 'nature' and femaleness that have consistently underpinned Western philosophy. One use Bacon identified for science and nature was the creation of an ideal society, the 'New Atlantis', in which the cultivation of gardens may be made 'by art, greater much than their nature; and their fruit greater and sweeter and of differing taste, smell, colour and figure from their nature' (Bacon, 1627: 1642). Parks and enclosures should be showcases for rare beasts and birds to view and experiment upon, and science experiments were designed to create optical devices, replicate sounds which might also be conveyed long distances, flying machines and underwater machines. It will not be lost on the reader that such ambitions have continued to inform Western science and, whilst the details have been modified and refined, today's Western science largely rests on the assumptions of Baconian philosophy: empiricism, objectivity and the control and utilisation of nature as a resource.[1]

Such a utopia was to be the creation of male experimenters 'besides a great number of servants and attendants, men and women' (ibid.: 1649). This consignment of women to the class of 'servants and attendants' typifies the period and has perpetuated the practice of science through to the present. Bacon, himself a powerful shaper of seventeeth-century social ideas through his essays and role as Lord Chancellor of England, wrote of wives that they 'are young men's mistresses, companions for middle age and old men's nurses' (1625: 1644), firmly entrenching the division of labour between men in the public sphere and women in the private or domestic sphere. Not all women, however, complied with this scheme, although they were often required to adopt aliases or rely on sympathetic male relatives, as the following section will show.

Female and male roles in the practice of science

Margaret Wertheim (1997) argues that women's exclusion from formal intellectual life became more entrenched with an increasing emphasis on 'expertise'. As knowledge became more formalised through the evolution of 'the academy' from cathedral schools to universities, and through the establishment of elite scientific bodies such as the French Academy of Sciences (founded in 1666, first admitted women in 1979) and the English Royal Society (founded in 1660, first admitted women in 1945), women were marginalised from intellectual life.

Thus, women whose intellect challenged the Aristotelian assumption of inferiority were denied the training now used to legitimise science. As Christine of Pisan (1364–1430) argued in her *Book of the City of Ladies*:

'If it were the custom to put the little maidens to school and . . . to learn the sciences as they do the man children . . . they should learn as perfectly' (quoted in Wertheim, 1997: 56).

High-ranking women used their social position to gain access to intellectual debate. For example, Margaret Cavendish, Duchess of Newcastle, used her husband's position to establish contact with Hobbes and Descartes who gave high regard to her books on natural philosophy. Similarly, Elizabeth of Bohemia corresponded with Descartes, and Queen Christina of Sweden employed him in order to discuss philosophy and science (Wertheim, 1997: 102), whilst Georgiana, Duchess of Devonshire formed a lifelong friendship in the 1790s with the scientist, Sir Charles Blagden. With his encouragement, Georgiana became an 'amateur chemist and mineralogist of note, later endowing Chatsworth with a collection of stones and minerals of museum quality' (Foreman, 1999).

Nonetheless, for intelligent but less high-born women, often their only channel for inquiry was through a sympathetic male family member. For example, Wertheim shows how women astronomers in the fourteenth and fifteenth centuries assisted their husbands, brothers and sons in order to pursue their intellectual interests, whilst Cheryl McEwan (1997) charts how, in eighteenth- and nineteenth-century Britain, women who were interested in the physical sciences had to satisfy their curiosity through assisting their brothers and fathers. Laboratory work was considered a legitimate pursuit for women, provided it focused on mechanical and menial pursuits, rather than on metaphysical contemplation, since the laboratory resembled a kitchen and therefore was likely to encourage domesticity. Books and periodicals were produced for women, for example, the *Free Thinker* (1718–1721) which published articles 'for the "philosophical girl" who "did not aspire to masculine virtues but was above female capriciousness"' (Merchant, 1983: 273). The one science in which women were encouraged, botany, was seen as being 'safe' since it could be studied in gardens and utilised such 'feminine' skills as drawing and painting (McEwan, 1997). This field has been distinguished by a number of women, the more noted of which have travelled widely in their search for botanical specimens. Their stories tell not only of their skill and independence, but of the release such activities gave them from their heavily prescribed lives, for example, Marianne North (see Figure 2.1) who was an artist and botanist in the nineteenth century whose health and well being dramatically improved whenever she spent time overseas. Botany is also linked with herbs and their curative and preventative properties with which women have long been associated,

Figure 2.1 *Marianne North (photo)*
Source: Courtesy of the Royal Botanical Gardens, Kew.

particularly before 'modern' medicine. Chapter 4 will explore the relationship between gender and the medical profession.

This division of academic labour has left its legacy in late twentieth-century Western science, with women constituting 16 per cent of tenured academic staff in UK science departments (PRISM, 1995) and only 3 per cent in UK Physics departments (Hall, 1997). Research by the Geological Society (the professional body for geologists) has reported that 12 per cent of its membership are female and only 4 per cent of chartered geologists. The research found that women geologists felt that the society was an 'old boys' club' in which women often felt patronised. In addition, 30 per cent of women geologists claimed that their careers had suffered through being female and similar proportions believed that they had been victims of sex discrimination, had reported minor sexual harassment (3 per cent had reported serious sexual harassment) and felt excluded by their male peers (Nield, 1998).

Even in the 'softer' science of geography (sometimes not considered a science at all, but a social science or humanity) only 6.9 per cent of full-time university lecturers were women in 1988, although one-quarter of the Research Division of the Royal Geographical Society (with the Institute of British Geographers) were women (Rose, 1993: 1).[2] A survey conducted in 1996 established that 22 per cent of staff in UK geography departments in the UK were women: 23 per cent of human geography staff and 21 per cent of physical geography staff. Two departments had no women geography staff, 21 had no female physical geographer and 14 had no female human geographer. Forty per cent of postgraduate research students were women: 45 per cent of human geographers and 37 per cent of physical geographers (Dumayne-Peaty and Wellens, 1998).[3] A pyramidal structure occurs in all disciplines whereby the majority of women hold posts at the technician level and decline in proportion and in number to the level of professor. Even for those women who do achieve academic prominence, wider recognition is limited. As Box 2.2 shows, from 1901 until 1995 only 9 of 409 Nobel prizes for science have been awarded to women (Wertheim, 1997: 236). This striking disparity in the scientific academy, at a time when the number of women undergraduates is overtaking men overall, raises the question of why the sciences have appeared to be so gender-selective and what the implications of this gender bias are on the practice of science and its impact on the environment.

Even in Sweden, named by the United Nations as the leading country in the world with regard to equal opportunities, discrimination against

Box 2.2

Women Nobel Laureates in science, 1901–1995

Name	Date	Field
Marie Sklodowska-Curie	1903	Physics
Marie Sklodowska-Curie	1911	Chemistry
Irene Joliot-Curie	1935	Chemistry
Gerty Raduitz-Cori	1947	Medicine
Maria Goepport-Mayer	1963	Physics
Dorothy Crowfoot-Hodgkin	1964	Chemistry
Rosalyn Sussman-Yalow	1977	Medicine
Barbara McClintock	1983	Medicine
Rita Levi-Montalcini	1986	Medicine
Gertrude B Elion	1988	Medicine

Source: Wertheim, 1997

women scientists has been established by research cited by Susan Greenfield. Greenfield, a neuro-physicist at the University of Oxford, is a persistent critic of equal opportunities in higher education and in Box 2.3 her impassioned reading of the Swedish report is reproduced from her bi-weekly column in the *Independent on Sunday*.

Brian Easlea suggests that in order to be a successful scientist, one needs ambition, competitiveness, aggression and 'a healthy sense of ego' (Easlea, 1986: 135, quoting Pagels). The practice of working long hours in often remote laboratories militates against women's participation given the current social framework in which women, even if professionally qualified and salaried, are primarily responsible for domestic chores and childcare, as Chapter 5 on the family will show. Traweek's research into particle physicists has established that 95 per cent are men whose wives stay home and care for 'hearth and husband'. Almost all the wives interviewed 'felt honoured to be playing this supportive role to men at the forefront of such important scientific knowledge'. The 5 per cent of particle physicists who were women tended to be married to other physicists (Traweek in Wertheim, 1997: 236).

As well as the approach to scientific work, the subjects prioritised by science are the product of arguably gendered interests. Easlea further

Box 2.3

Susan Greenfield

UNDER THE MICROSCOPE

BY SUSAN GREENFIELD

Unbelievable

IT IS VERY rare that I get truly angry. However two Swedish scientists have managed to turn my knuckles rage-white with a ground-breaking paper published in *Nature* at the end of May. It was not that they were reporting anything particularly surprising: everyone I have spoken to since has effected little surprise. Rather, the shock was to see definitive and objective proof that female scientists are considered less competent than their male counterparts.

Christine Wenneras, a microbiologist, and Agnes Wold, an immunologist, were attempting to discover why women scientists are about half as successful as men in obtaining fellowships from the Swedish Medical Council. Is female performance really so inherently feeble next to men on such a conspicuous, generic scale? No. The problem is not with how good a scientist an applicant is, but rather how they are perceived. Wenneras and Wold used a robust and objective measure to ensure that they could compare the success of men and women, who, to date, had been equally productive. They calculated an "impact" score based not only on how many scientific papers the applicant had published, but also on how many times that paper had been cited by other scientists; in other words, how important the findings were to the rest of the scientific community.

The results were staggering, and disgraceful. Women with the highest impact scores were rated, in terms of subjective views of scientific competence, only slightly above the very worst of the men. For a female scientist to be judged as competent as a man, she had to exceed his productivity by some 20 research papers. Taking my own field, somewhat arbitrarily, as representative, 20 more papers would amount to about 10 years more work, if, of course, one was lucky enough to get the funding. The only other factor that counted in mitigating one's lack of a Y chromosome was to be in cahoots with at least one of the reviewing panel. So as if sexism in the Swedish MRC is not bad

continued . . .

enough, the situation is compounded by an old-boy system, again already suspected to be rife in the peer review system, but never before actually proven.

My anger on reading these figures was rooted in frustration. Earlier suspicions this time could not be tempered by the view that one was, at least in part, hysterical, strident, or paranoid. It is simply wrong. But brute prejudice is as hard to annihilate as it is ugly. What can we actually do?

The problem surely extends into daily laboratory life as well. If our "peers" perceive us as incompetent when on a panel, then the prejudice will come into play at the bench. And the distorted perception will not just go away because those of us with access to the media say that it should.

This study was performed in Sweden and some might argue it is different over here. Too right. We do not have the freedom of information legislation Sweden has, so records cannot readily be obtained. Indeed, even in Sweden, Wenneras and Wold could

only obtain the relevant data after a court battle. How refreshing if British funding bodies voluntarily made their data available for analogous scrutiny. The story is made more dismal as Sweden was named by the UN as the leading country in the world with respect to equal opportunities.

When I gave a talk recently to a women's group at an Oxford college, the discussion threw up all number of seemingly trivial incidents – the backhanded comment, the dismissive body language, the ignoring of a comment. All in themselves too silly to protest about, yet cumulatively eroding confidence. Perhaps it is here, in this most basic of grass roots, that we have to take a stand against the assumption that unless we have worked for 10 further years, we are worse than most men. The authors' conclusion was that unless things change in science, ". . . a large pool of promising talent will be wasted." Quite an understatement

Source: Courtesy of the *Independent on Sunday*

argues that the Baconian/Enlightenment emphasis on objectivity, rationality and control alienates women, reinforcing the natural sciences as 'masculinist'. Any subject whose founder (Francis Bacon) enjoins men to stop making war on each other so they may make a 'united scientific assault on 'female' nature' is unlikely to appeal to more than a minority of females (Easlea, 1986: 149). The current military emphasis of physics is known to discourage female participation, women being less likely to want to pursue work in research that is defence-related or involves animal experimentation (PRISM, 1995). Wertheim's central argument in her book *Pythagoras' Trousers* is that physics has been devoted to a central quasi-religious 'quest' in its search for the origins of the universe. As in most religious orders, women have been considered ill equipped to rise to the transcendental challenge. She further argues that this quest, currently manifested as Unified Field Theory, is largely irrelevant compared to pressing problems such as pollution, over-population, starvation, land degradation, deforestation and biodiversity loss. If more women were involved in physics, she argues, perhaps the goals and expenditures of physics would be re-evaluated. (For example, before its recent cancellation, the cost of the US Superconducting Supercollider[4] was proposed to be $10 billion, possibly even $13 billion. Wertheim, 1997: 237–238).

Geography, a discipline centrally involved in the study of the environment, has also been powerfully shaped by gendered practice. Gillian Rose has demonstrated how the nature/culture dualism is deeply embedded in the subject, but argues that the relationship between the observer and landscape represents more than impartial scientific objectification. She shows how *the gaze*, which is a way in which the powerful regard their object through distance and dominance, is also constructed by an aesthetic and sensual appreciation of landscape (and people who populate these landscapes) which can actually undermine this *objectivity*, although this is rarely acknowldeged.[5] The nature or landscape which is regarded has often been sexualised as female by descriptions such as 'Mother Earth' or the 'Sirens which seduce the gazer' (Rose, 1993: 72). Rose projects the discipline of geography as quite masculinist, not only for this reason, but also because it replicates a masculinist culture of drinking and ribaldry with which students, as well as staff, are expected to conform (Rose, 1993: 68). Sarah Maguire's research reported in a special issue of *Area* dedicated to gender and physical geography has queried the nature of physically strenuous fieldwork which was found to discourage some female students (Maguire, 1998).[6]

These points raise a number of questions. What is it about science that requires its practitioners to be competitive and aggressive? Who chooses which objects are to be studied? Who decides the appropriate methodology?

The gendering of science

One suggestion is that the practitioners of science have, consciously or unconsciously, shaped their discipline. The institutions which variously employ and fund scientists and their research are socially constructed, and staffed by the elite of that society: white, male, professional. Consequently, gender relations (as well as class and race relations) are likely to have a bearing on the answers to the above questions. Sandra Harding argues that if women's concerns were the starting point for scientific problems, rather than men's, ecological research may well be preferred over military, or research with a social benefit preferred over that which serves solely industrial interests (1990: 109).

Since the 1980s, there has been an increasing amount of evidence to suggest that the practice of 'modern, Western science' has, indeed, incorporated a significant gender bias which affects the nature and quality of scientific research. Donna Haraway (1978) pioneered this research by comparing projects investigating primate behaviour with a view to explaining the evolution of human social behaviour. The accepted explanation of human behaviour which held sway in the early and mid twentieth century stressed the importance of aggression and competition, characteristics which have been thought to be necessary in order to hunt. This idea was based on research into male primates by Sherwood Washburn and David Hamburg which was then applied to human society to try to account for a seemingly human male attraction to regulated fighting, torturing and killing (Haraway, 1978: 62). Thus (male and female) behaviour are explained by observing the traits of one sex.

Conversely and, in the scientific community more controversially, Nancy Tanner and Adrienne Zihlman have focused on the study of the female primate, arguing that this hunting thesis has 'largely ignored the behaviour and social activity of one of the two sexes', rendering it rather suspect (in Haraway, 1978). Noting the development of gathering of food, chimpanzees' use of tools (most frequently by the female) and a flexible social structure, they suggest that rather than early human behaviour being characterised by aggression and competition, it was equally focused on cooperation and sociability. Male–female cooperation was needed to engage in the difficult task of human child-rearing and males extended these relationships to strangers, forming linguistic communities. Likewise, Harding cites anthropological research which suggests that 'woman-the-gatherer' is a more plausible explanation of evolution (1986: 98).

Where male researchers focused on a study of male primates in an attempt to explain food provision and social organisation, they established that such provision and organisation necessitated aggression and dominance. Conversely, where women researchers focused on female primates to explain the same processes, they found that this required the opposite skills of cooperation and sociability. This stresses how the researchers' point of view (in this case, being male or female) can affect the design and results of the research. This much-simplified synopsis of Haraway's reflection on primate research illustrates quite clearly how gender has a major impact on the framing of scientific questions and, consequently, their answers. This impact has also been felt in the Social Sciences. Jane Ussher gives examples of psychological research into theories of moral development and of ageing, both of which were conducted entirely on samples of boys, although the results were generalised to both sexes (Ussher, 1997). Haraway's undermining of the success story of aggression and competitive behaviour is echoed by Evelyn Fox Keller (1991) who challenges the notion of 'the selfish gene' popularised by Richard Dawkins (1976). This interpretation, of course, has a long history, dating back to Darwin's thesis of survival of the fittest. Keller suggests that there is increasing interest in mutualist relationships which is eroding the previously near universal belief that humans are genetically predisposed to selfishness for purposes of survival.

Keller also shows how it is not just the setting of questions and the arrival at answers which is gendered, but how this extends to the methods of analysis used. The point that I would like you to bear in mind here, is not whether this way of thinking is 'right', but that it is at least equally valid as the more masculine way of thinking, which is usually valued more highly. In Barbara McClintock's case, her approach led her to a breakthrough in cytology: in Box 2.4, Keller illustrates how Barbara McClintock revolutionised genetics by using values usually described as 'feminine' to inform her research into corn genes. This is not to say that only women scientists could conceptualise 'mutualist' theories of evolution, nor that only men scientists would develop 'competitive' theories. However, these examples highlight how certain values *may* be affected by gender. It is also interesting to contemplate how well publicised Richard Dawkins and his 'selfish gene' theory is compared to Barbara McClintock, whom I had certainly not heard of before I started my research for this book.

As a final example of how the practice of science is gendered, Emily Martin's work on the metaphors of reproduction will be used to show the powerful effect language has on interpretation (Martin, 1991). Martin illustrates how persistently images of the 'active' sperm and the 'passive' egg are portrayed, even though current research suggests that mutual action is involved (ibid.: 110). She cites examples from research articles and textbooks which eulogise the sperm – protected by its 'vestments', having a 'crown' and accompanied by 'attendant cells', whilst the egg awaits rescue from its infertility like a 'sleeping beauty'. Such descriptions strongly reinforce male and female social roles (and, since these images appear in textbooks, reinforce the way in which students are

Box 2.4

Barbara McClintock

Evelyn Fox Keller, herself a physicist, has documented the life of the inspired American geneticist, Barbara McClintock who was denied many academic opportunties because of her sex and for whom recognition was much delayed because of her emphasis on the whole organism or, as McClintock herself articulated it 'above all, one must have a feeling for the organism'.

Barbara McClintock pioneered work into genetic inheritance through the use of maize. Although this research was commonly conducted on the fruit fly because of its fast reproductive cycle, McClintock favoured maize because its variegated kernal colours were, she felt, better suited to microscopic study. Assisted by Harriet Creighton, she established the chromosomal basis of genetics although her focus on the whole organism and her respect for 'oneness' (or the holistic nature) of things was reflected in her often very complex explanations which found little favour with the molecular biologists who dominated genetics in the mid-century. McClintock's theory that the cell's environment was critical to its performance was eventually recognised by the 1970s, culminating in her Nobel Prize in 1983, at the age of eighty-one.

Her groundbreaking research was pursued against all odds. She enrolled in the Cornell College of Agriculture, New York, a college known to be hospitable to women at the time. However, her chosen PhD field of study, genetics, could only be taken in the Department of Botany as the Plant Breeding and Genetics Department did not enrol women. The accepted route for women was to become scientific workers (laboratory assistants) or teachers; a career as a reaserch scientist was simply not open to them, nor were faculty positions in universities. McClintock's female students and colleagues generally went on to teach in women's colleges (as did Harriet Creighton), or to work in their husband's laboratory (McClintock never married).

Until she moved to the Cold Spring Harbour Laboratory on Long Island, New York, McClintock's career was a series of temporary fellowships. By 1933, '[h]ere was McClintock . . . with the best of credentials, backed by the giants of the field and unable to get a job'. 'The best trained and most able person in this country on the cytology of maize genetics' was considered virtually unemployable. In 1941 she secured the first of many Carnegie research posts based at Cold Spring Harbour where she stayed until she retired.

Keller stresses McClintock's holistic approach as being an essential ingredient of her insight and shows how this leads to a much fuller understanding of how scientific approaches impact on the environment.

> We've been spoiling the environment just dreadfully and thinking we were fine, because we were using the wrong techniques of science. Then it turns into technology, and its slapping us back because we didn't think it through. We were making assumptions we had no right to make. From the point of view of how the

> whole thing actually worked, we knew how part of it worked . . . We didn't even inquire, didn't even see how the rest was going on. All these other things [Love Canal and the acidification of the Adirondack Lakes, USA] were happening and we didn't see it . . . Technology is fine, but the scientists and engineers only partially think through their problems. They solve certain aspects, but not the total and as a consequence it is slapping us back in the face very hard.
>
> (From an interview with Barbara McClintock)
>
> *Source*: Keller, 1983: 198, 72, 74, 205–6.

socialised). Martin also castigates the tendency to denigrate the functions of the female body. So, whilst the sperm is described by positive, healthy, images of speed, velocity and power, ova are usually described as degenerating from birth, menstruation is construed as failed production and an ovary of a middle-aged women as 'a scarred, battered organ' (ibid.: 104–105).

Other research has shown how women's bodies have been considered 'different' from men's and how women have been considered legitimate targets for contraceptive experimentation (Oudshoorn, 1996: 158) particularly with respect to the 'problem' of over-population as defined by Western scientists (Harding, 1986: 77). Women in the South have been particularly vulnerable to such experimentation, as have ethnic minority women in the West, as Chapter 4 will show. Older women are increasingly being seen by the pharmaceutical industry as effective targets, whereby menopausal and post-menopausal symptoms are conceived of as 'problems' which need to be treated, by medical interventions such as oestrogen-based hormone replacement therapy, by a society which values youth over age and which denigrates 'crones' and 'hysterical' women (Greer, 1991; Leyson, 1996). Thus, it is not difficult to see how medical and scientific research is problematising both female fertility and the female aging process.[7] These research initiatives are seen as acceptable, or even welcome, because of society's continuous exposure to a language of medical science which normalises the concepts of fertility (in the developing world) and menopause as problems. These points concerning society's treatment of the body will be returned to in Chapter 4, whilst Chapter 7 will develop the global context in which some of these relationships are played out.

So, if the practice of science is a product of a set of gender relations and if that is producing biased, or only partial, results (what Mary Tiles refers to as a 'Science of Mars' (Tiles, 1987)), is it possible for science to become less gender-biased, or gender-neutral? The final section of this chapter considers the broad spectrum of feminist approaches to the search for knowledge (or methodologies), which its proponents argue can make science more effective.

Redressing the gender bias in science

Whilst practitioners of science have claimed objectivity in their abstract endeavours, the arguments earlier in this chapter, along with the examples and case studies, suggest that it is inconceivable that scientists are able to step outside their social, political, economic and moral framework when they enter the laboratory, field or study. Every part of the research process is shaped by these frameworks; from deciding the research question; the methods used and the theoretical perspectives employed.

Harding (1994) argues that feminist theoretical perspectives (such as those which are outlined below) are more successful in highlighting the androcentricism (that is, male-centredness) of traditional science, showing how the observer/researcher cannot stand outside his (sic) research. Whilst the issue of subjectivity is most closely associated with feminist theoretical approaches, Keller (1982) suggests that the realisation that we cannot completely separate ourselves from what we research was developed amongst a number of men and women throughout the twentieth century, providing the basis for the current feminist critique. As Dorothy Smith stated in 1974, 'the only way of knowing a socially constructed world is knowing it from within. We can never stand outside it' (Smith, 1974: 24). The rest of this section introduces methodologies which differ from traditional science. These are summarised in Box 2.5 alongside 'Scientific Method' to give a quick overview of their values, objectives and limitations.

Three approaches offer a way of changing science and knowledge acquisition in general. First, *feminist empiricism* which strives to introduce the female into science as part of what should be studied, the student and researcher, and the practice, utilising skills usually identified as more 'feminine' such as cooperation and teamwork, to challenge the aggressive, dominant style. This approach is an 'additive' one, seeking to adapt current practice rather than revolutionise it. Evelyn Fox Keller is one of the main proponents of this approach arguing that objectivity is valuable as an ideal and that the inclusion of more women in science, and

Box 2.5

Approaches to scientific inquiry

Methodology	Values/Objectives	Problems
Scientific method	Logical	Logic is defined socially and may vary through time.
	Objective	The 'knower' cannot completely separate from what is to be known.
	Empiricist	What is observed may be biased.
	Value-free	Science is a product of its society and is therefore informed by its values.
Feminist empiricism	Embraces feminist values.	
	Ensures what is observed includes male and female, all races, classes, etc.	
	Objectivity is an ideal.	
	Include more women in scientific practice.	Danger of women becoming 'honorary men', if science itself is unchanged.
		All the above characteristics may not challenge the underlying method.
Situated knowledge	Recognises and acknowledges the position in which the researcher is located.	Danger of having so much diversity, it is impossible to make any generalisations. How is positionality controlled for?
Feminist standpoint	Oppressed people's viewpoint is more valid as they are more grounded in material reality.	Whose viewpoint do we use? Are women more oppressed than, say, colonised people?
	Only research seeking to eliminate repression can hope to decrease subjectivity.	

continued . . .

Box 2.5 continued

Methodology	Values/Objectives	Problems
		Ignores differences within group and potential conflicts. Unrealistic to look for a unified explanation.
Postmodern	Respects difference.	
	Impossible to concentrate diversity into a unified 'story'.	Unlikely to be able to challenge the status quo, therefore remains marginal and does not advance the political project.

the use of such approaches (which may, it should be pointed out, also be practised by men, just as women may practice more 'masculine' science) make that ideal more achievable. That is, it reduces the bias (as the earlier example of primates research cited by Donna Haraway illustrated) and permits a more holistic and balanced view of the problem. In valuing the input of women and feminine approaches, science becomes more inclusive by ensuring, for example, that large sections of the population are not left out of experiments, as the work of Haraway and Ussher showed earlier. Rather than biasing science towards women, this approach claims more balanced and equitable results. In turn, the areas identified for research may change and, by using alternative methodologies, such as those illustrated in the case study of Barbara McClintock, new breakthroughs may be made.

Second, Donna Haraway (1988: 254) stresses that a 'feminist objectivity' means recognising where you are seeing from and acknowledging that your knowledge can only ever be partial. This *situated knowledge* is also given the name of 'positionality' since it is the position from where you see which will determine what you see (see Kim England, 1994, for a discussion of this). Since everyone who is engaged in research is influenced by a range of social factors (such as their sex, gender, age, social background, ablebodiedness and so on) these need to be acknowledged to avoid concealing our biases. This approach holds that such revelation renders our research more honest as, even though researchers will try to correct for any bias in setting up and conducting their research, it is impossible to eliminate it entirely.

The third approach is most radical and originates from the social sciences; it is known as *standpoint theory*. Nancy Hartsock (1987) explains that the traditional perspective of research has been the vantage point of the privileged in society. In Western society, this means that science is 'done' by white, elite, educated males. In order to accomplish this work men/scientists are less in direct contact with processes of reproduction which encompass all those practices which make their work possible (from the cooking of their meals, laundering their clothes, servicing their cars, driving trains to transport them to conferences, cleaning the offices and laboratories in which they work and typing the fruits of this work) and so, as Hartsock (1987) and Harding (1994) argue, they are actually more removed from what may be termed real life and social relations. Hartsock goes on to suggest that these abstract researchers, whilst being out of touch with those practices itemised above, also have a vested interest in maintaining the status quo. In order to do so, they will probably need to reveal their research results selectively.

Hartsock, Harding and other proponents of standpoint theory argue that research can much more effectively be done from a position which has a more solid grounding in the material (as opposed to abstract) world, and from an experience of inequality; the two of these are almost always interlinked.

A specifically feminist standpoint, then (where the 'knower' is a woman involved in the material world and subservient to men), it is argued, is more likely to reveal the 'reality' of social relations. If a feminist standpoint is taken, therefore, a greater understanding of social relations (or social-nature relations) is likely and the questions the researcher is asking are likely to be different, grounded as they will be, in her experience. After all, who is to say that her priorities are any less important than those identified by the 'impartial' scientist? Whilst standpoint theory is more commonly associated with the feminist standpoint, its description will suggest that any position occupied by an oppressed and materially grounded group (e.g. colonised people, ethnic minority groups, disabled people) is a legitimate and valid standpoint.

Responding to postmodernist claims that social fragmentation makes it impossible to generalise about all women (or, for that matter, all disabled people, ethnic groups, children or men), Harding stresses the value of what she terms a 'successor science'. By this she means a unified explanation which should replace or succeed the current Western claim that science is neutral. Whilst recognising inter-group differences, she argues that only a common project espousing the participatory values of anti-sexism, anti-racism and anti-classism has the potential to reduce the subjectivity embedded in current scientific practice (Harding, 1986: 49).

Environmental research is also governed by the codes of claimed abstract objectivity, be it research into pesticides, global warming or acid rain. The feminist standpoint offers a useful critique of this in that women, more grounded in material reality or, put more prosaically, more accustomed than men to wash, cook, clean and care for the health of family members, are most likely to experience 'real' social-environment relations. By what Harding and Mary Mellor call their 'embodiedment' (that is, being more closely in touch with their bodies because of their domestic and child-bearing and rearing activities) they are exposed to the results of environmental damage more directly than men, particularly those who are abstractly 'making science' (Harding, 1986; Mellor, 1996). The eco-feminist arguments in the next chapter will explore the validity of privileging women's knowledge on the basis of both their relationship with the environment and on the unequal relationship between men and women.

Summary

This chapter has set the historical context for the pursuit of scientific knowledge practised in Western society. It argues that the architects of this Scientific Method were strongly influenced by, and worked within, a prevailing set of social norms which dictated that only men should engage in scientific research and that participation in this was confined to an elite qualified few. This has helped to shape both the knowledge and the way in which it is practised which currently underpins Western society. Examples have been shown of how research that is claimed to be objective and neutral is, in fact, heavily biased and the chapter concludes with suggestions as to how science could be practiced more fairly by acknowledging its bias, and opening it up to people who have traditionally been excluded from its practice.

In addition to questions posed within this chapter – such as who chooses the objects to be studied and the methodologies to be used? why do scientists need to be competitive? – you might also like to consider the kinds of questions and problems addressed in your own course and whether these reflect particular masculinist (or feminist) positions. At the same time, reflect on the gender balance of your own department: of undergraduate and postgraduate students, technicians, junior and senior academic staff, full-time and part-time staff. Finally, you might also like to reflect upon your own fieldwork experience. How far does this mirror the masculinist practice suggested by Gillian Rose and Sarah MacGuire? And, if it does, how do you feel about this?

Discussion questions

1 Thinking of the department in which you are studying, what is the gender balance of undergraduate students, postgraduate students, technicians, lecturers, readers and professors?

2 How many of the women Nobel Laureates do you recognise? Why do you think the number is so low?

3 Consider your own fieldwork experience. To what extent does this conform to the experience reported by Gillian Rose and Sarah Maguire?

4 Considering a piece of research that you are undertaking, and the courses that you are enrolled in, what are the underlying assumptions here, what influenced the posing of the questions and choice of subject and methodology?

Further reading

Keller, E. F. and Longino, H.E. (eds) (1996) *Feminism and Science*, Oxford: Oxford University Press. This is a collection of previously published papers on how science has been gendered. Both the editors and their contributors are distinguished writers in the field and the book provides an opportunity to dip into debates in this area through the past twenty years.

Lykke, N. and Braidotti, R. (eds) (1996) *Between Monsters, Goddesses and Cyborgs – Feminist Confrontations with Science, Medicine and Cyberspace*, London: Zed Books. This is also an edited collection of papers which cover contemporary debates such as the medicalisation of women's experiences.

Pepper, D. (1996) *Modern Environmentalism: An Introduction*, London: Routledge. This is a good introduction to the background to the modern environmental movement, tracing its history back to medieval times.

Wertheim, M. (1997) *Pythagoras's Trousers: God, Physics and Gender Wars*, London: Fourth Estate. A very readable book written by a science journalist. Her main argument is that the pursuit of a science is equivalent to a religion and that its scientists are like priests who have followed a very particular line of inquiry from which non-initiates have been excluded.

Notes

1 Having said this, quantum science has radically changed the way in which scientists view the smallest (micro) scale and the largest (macro) scale, particularly through ideas such as relativity which holds that objects do change depending on where they are viewed from. Quantum physics now disputes the possibility of material stability at the micro scale, in that protons manifest both as matter and as energy waves (see Capra, 1976 for an explanation of this). One interesting aspect of this is that these developments have not yet been adapted by mainstream *social* science in the way that it adapted Newtonian, mechanistic, science in the twentieth century.

2 The Royal Geographical Society with the Institute of British Geographers is the UK's leading geographical organisastion to which most academic geographers belong.

3 The response rate for this survey was 74 per cent. It does not distinguish between part-time and full-time staff.

4 The Superconducting Supercollider was designed to attempt to distinguish between energy-waves and matter in protons, see note 1.

5 This argument has been put most powerfully, with regard to the way in which Europeans regarded the Orient from the seventeenth century, by Edward Said (1975).

6 This is not meant to generalise levels of fitness between men and women; the main complaint seemed to be, not that some physical effort was involved, but that the pace of this activity was set by the men and that the exercise element (as opposed to the skills of observation) was seen as a heroic endeavour. Maguire quotes Stoddart as claiming that 'fieldwork is as much about the physical challenge as it is to do with furthering our knowledge of geography' (Stoddart, 1986 in Maguire, 1998).

7 Whilst fear of ageing and its denigration is not unique to women, the male ageing process tends to be more subtle and is not subject to the same interventions as those for menopausal women. Having said that, the recent public discussion over Viagra (a drug supposed to counter impotence) reveals the premium society puts on men's virility.

3 Explaining gender–environment relationships

Key words: ecological feminism; patriarchy; social eco-feminism; cultural eco-feminism; ethic of care/partnership

- Cultural eco-feminism
- Social eco-feminism
- Critiques and ideals

Chapter 2 has outlined the relationship between gender and science which has helped to create, particularly Western, societies' relationship with nature. This is evident from early Greek thought, which sought to impose a presumed masculine intellectual order on to what was considered to be a feminine chaotic nature, through to colonial settlements in which 'virgin female nature succumbed to the male plough' (Frank Norris in Merchant, 1996: 44). This scenario in which man (sic) fears and seeks to tame nature is still in evidence today, as examples in this book make clear.

In this chapter I want to look at a particular set of arguments which offer a critique of this Western world-view. These arguments can be broadly captured under the banner of ecological feminism (often shortened to eco-feminism). As the term suggests, this movement owes much to both the ecological and feminist movements of the 1960s and 1970s, although its synthesis argues that ecologism without feminism and vice versa is not enough to reverse the social and natural damage caused by Western society. For example, the liberal feminist movement has traditionally advocated equal rights for women as are accorded to men. However, without addressing ecological concerns, this would result in the dominant society increasing the damage done to nature, as more women would have access to a lifestyle which places a burden on nature. Also, some of the demands of liberal feminism for equal opportunities, for example, women's access to reproductive technology, can have negative effects on the environment and other women, as Chapter 4 will discuss. On the other hand, an environmental movement which does not address issues of equity (between women and men, as well as between other groups) can be accused of eco-fascism, where 'nature' is valued more highly than certain groups in society (see accusations by social ecologists, for example, such as Murray Bookchin and Janet Biehl cited later in this chapter). The most extreme form of this socially bereft form of 'environmentalism' found expression in German Nazism.

Ecological feminism believes that the patriarchal nature of Western society is to blame for the dominance of women by men and of nature by society, particularly through capitalism (though not exclusively: pre-capitalist societies have exhibited patriarchal tendencies, and still do in some cases) which, according to Ariel Salleh (1995: 23) appears as a modern form of patriarchal relations. Box 3.1 explains how patriarchy is built on the domination of one group by another. It is important, however, to be aware that patriarchy is a conceptual idea which organises and explains our experience: not all men dominate, and not all women are dominated, to the same degree. Moreover, there are other groups of people who have been subjugated by Western society and persistent inequalities exist not only between women and men but also between the colonised and colonisers, people of colour and a white elite, children and adults and wage labourers and owners/controllers of capital. However, since this book is concerned with gender, it is the triangular relationship between men, women and nature which this chapter addresses.

Box 3.1

Definition of patriarchy

This is the systematic, structural and unjustified domination of women by men. Patriarchy consists of those institutions (for example, policies, practices, positions, offices, roles and expectations) and behaviours which give privilege (higher status, value, prestige) and power to males or what historically is defined as the male gender. These institutions and behaviours constitute a sexist conceptual framework which in turn sustains and legitimises them. At the heart of patriarchy is the maintenance and justification of male gender privilege and power.

Source: adapted from Karen Warren, 1994: 181.

Eco-feminism was coined as a term in 1974 when Frenchwoman
Françoise d'Eubonne called for an ecological revolution to be led by
women in order to save planet Earth. Eco-feminism has evolved both as
an analysis of society–nature relations and as a prescription for how these
relations can be transformed. Broadly, these analyses fall within two
areas: cultural eco-feminism and social eco-feminism. Cultural eco-
feminism identifies a powerful and positive link between women and
nature, particularly through such female reproductive functions as
childbirth and menstruation. This connection between women and nature
is used to argue that women are better placed than men as advocates of
nature. Social eco-feminism argues that because women and nature have
both been subjugated by a society dominated by men, women, through
the roles they play, are in a better position than men to speak for nature,
because of this shared experience of domination. Social eco-feminists
contest that there is anything more natural in a woman's body than in a
man's and disagree with cultural eco-feminists' belief that there is
something which consititutes a woman's 'essence'.

Just as Chapter 2 placed particular aspects of Western thought in their
social and historical context, eco-feminism also needs to be seen as a
product of the social, political, environmental and historical context in
which the writers are working. Both cultural and social eco-feminism are
responses to the dualistic forms of thinking that have underpinned
traditional Western thought. Each position will be reviewed in turn,
before exploring more recent critiques of cultural and social eco-
feminism which argue for a dissolution of dualistic thinking altogther and
a move towards greater partnership between human beings.

Cultural eco-feminism

Chapter 2 illustrated the dualistic philosophy that has underpinned
Western thought. In this, not only are characteristics conceived of as
opposites (masculine opposing feminine, natural opposing abstract, for
example), but one set of characteristics (masculine, abstract, logical) are
elevated in importance over others (feminine, natural, chaotic). This is
used to justify the power that one group of people has consistently held
over others. Cultural eco-feminists attempt to reverse this hierarchy,
demonstrating the positive side of those characteristics previously held to
be inferior and stressing the importance of women–nature links to the
survival of nature.

Writers such as Mary Daly (1978) and Andrée Collard (1988) emphasise the physiological bond between women and nature through the capacity of women to carry, give birth to and nurse children. They argue that bodily processes, such as menstruation, bring women into regular contact with the processes of nature (for example, lunar and tidal rythms). But rather than using this experience to exclude women from public life, cultural eco-feminists argue that these very experiences enable women to make more sensitive decisions concerning, and construct more harmonious relationships between, human beings and between society and nature. Through their biological role, and the subsequent caring of children and the household, women develop an ethic of care which is thought to be essential to revaluing the relationship between humanity and nature.

Cultural eco-feminists (sometimes termed essentialist because of their belief in the biological essence of femaleness) promote their beliefs through a celebration of the feminine, particularly rituals of menarche,[1] childbirth and through the revival of goddess worship. Charlene Spretnak, for example, celebrated the myth of Gaia: Earth mother in her book 'Lost Goddesses of Early Greece', emphasising the powerful spirituality of ancient goddess traditions which revered the Earth (Merchant, 1996: 3). Maria Mies and Vandana Shiva refer to a spiritual eco-feminism which taps into women's ancient wisdom (such as knowledge of plants) and their sensual energy. In celebrating the naturalness of the body, humans' connection to the Earth is reinforced (Mies and Shiva, 1993: 16–17).

Such a perspective appeals to Gillian Rose (1993: 79) who values 'the most wonderful parts of these books' which identify 'places in which they [women] succeed in imagining a space in which women might really be free'. However, she ultimately rejects this interpretation of gender/environment relations as the section on critiques and ideals will discuss below.

Social eco-feminism

Rather than believing that women are closer to nature through their bodily functions (after all, all bodies are part of nature), this perspective argues that it is the social role ascribed to women which identifies them more closely with nature. In this interpretation, women's closeness to nature is seen as socially constructed, that is, a product of the role women have been socialised into through generations. Whilst only women may conceive and give birth to children, the caring and domestic role assigned to mothers *vis-à-vis* fathers explains the degree of embodiedness they experience. Mary Mellor, a British protagonist of social eco-feminism, explains 'embodiedness' as the process by which women retain intimacy with their bodies and the inputs bodies need for survival. On the other hand, by assigning mothers this caring/domestic role, fathers are able to distance themselves from processes which ensure their daily reproduction (for example, the growing and preparing of food, collection of water and cooking and heating fuel), relying on women to provide their bodily needs such as food, water, warmth and clothing. Chapter 2 has shown how such extrication from the sources of nuturing allows people to more easily abstract themselves from nature. As Val Plumwood (1993) has stated, reproduction and subsistence have been neglected in favour of production, a point which will be developed in Chapters 5 and 6.

Social eco-feminists argue that women, who, because of their social roles are less able to distance themselves from nature and who experience subjugation and discrimination based on their socially ascribed caring role, are able to share with nature a feeling of being dominated. This entitles women to speak up for nature against its domination. The first step towards ending the social domination of nature would be to dismantle other forms of domination, since social eco-feminists believe that the idea of dominating nature stems from the domination of human by human (Biehl in Merchant, 1996: 13). The standpoint theory explained in Chapter 2 is useful in understanding this idea, since it proposes that only the most oppressed groups can truly understand interpersonal (or society-nature) relationships, as oppressor groups have too much much vested in these to challenge them.

Social and cultural eco-feminism can be placed at each end of a philosophical continuum and many ecological feminists will incorporate aspects of each of these together with other ecological and/or social analyses. For example, Hallen's list of ten varieties of eco-feminism, displayed in Box 3. 2, draws variously on deep ecology, emancipation, humanitarianism and social philosophies.

Critiques and ideals

Many people, including ecologists and feminists, have problems with these interpretations of eco-feminism, on a number of grounds. Feminists

---

Box 3.2

Eco-feminism as a broad, diverse, world-wide movement

Unique eco-feminist approaches

- Liberal eco-feminists who seek reform from within existing political and economic structures.
- Radical eco-feminists who wish to dismantle those very structures through direct action.
- Cultural eco-feminists who focus on the cultural manifestations of the women–nature connection, earth-based spirituality, goddess religions, and witchcraft.
- Social eco-feminists who build on the social ecology movement of the American anarchist philosopher Murray Bookchin in an attempt to restructure hierarchical society into egalitarian, decentralised, bioregional communities.
- Socialist eco-feminists who draw on neo-Marxist philosophies to focus on the relationship between production and reproduction and on women's work in the continued biological and social reproduction of life on Earth.
- Ecological eco-feminists who strive to show the respects in which eco-feminism and the science of ecology (specifically eco-system ecology) share vital similarities.
- Deep ecological eco-feminists who draw on the work of the Norwegian philospher, Arne Naess, and strive to dismantle both anthropocentrism (human-centredness) and androcentrism (male-centredness).
- Critical or transformative eco-feminists who wish to transform the very categories of masculine and feminine and the divisive nature of dualistic rationality.
- Aboriginal or native eco-feminists who live close to nature, nuturing sacred lands and reconsecrating degraded spaces.
- Eco-feminists of the Third World who criticise maldevelopment in the First World and show us how women of colour may be in a privileged position because their minds are not yet colonised and because they do not profit from the oppression of others.

Source: Hallen (1994: 207).

---

have fought long and hard against the idea that women are expected to have a subservient domestic role because of their capacity for bearing children. Despite the fact that cultural eco-feminists stress the importance of revaluing domestic tasks and child-bearing and rearing, the biological or essentialising imperative is seen as retrogressive as it affirms and naturalises women's role as nurturer and life giver (Plumwood, 1993:

31). Gillian Rose (1993) voices another concern with essentialism which is: if we agree that women have an 'essential' quality, which is nuturing and sympathetic, is this then universally so for all women? If so, this ignores any differences between women, that is between white women and women of colour, between those with children and those with none, between women in the North and women in the South, and so on. She also questions whether, then, the argument for essentialising race can be resurrected and, since she rejects this, this leads her to reject all essentialist arguments. We might also consider, as an extension of the essentialist argument, whether men's nature is also universally fixed, which perhaps would acknowledge that male aggression and competitiveness cannot be changed. As the next chapter will explore, it is questionable whether such fixed qualities can be attributed to individual bodies, when it is difficult to say with any certainty that the body is a fixed and immobile concept.

Val Plumwood fears that this focus on women's proximity to nature can lead to burdening women with cleaning up the environment (as a kind of global super housewife), appealing to their feelings of guilt and motherly duty (ibid.: 23). Or, as Janet Biehl (1991: 26) has expressed it '[o]nce again women are being asked to take the fall – this time to save the planet' (Figure 3.1).

There is also some concern that adopting a dualistic eco-feminist analysis excludes all men. Carolyn Merchant, particularly, advocates a partnership ethic in which men and women enjoy a non-hierarchical partnership with nature (Merchant, 1996). This is a view shared by Karen Warren (1994) who prioritises interpersonal relationships of mutuality, care, reciprocity, friendship, appropriate trust and love. Plumwood specifically urges us to do away with dualistic thinking altogther. She argues that one side defines the other in this kind of thinking so that the feminine traits of caring, naturalness and embodiedness are themselves the result of domination. It is not, therefore, sufficiently liberating to elevate the dominated side. Besides, without the dominant abstract, logical, masculine, perhaps the dominated chaotic, natural and feminine may become something quite different. Rose, in her rejection of the essentialist, cultural eco-feminist position, suggests that the male/female dualism is, in fact, a male/male-interpretation-of-what-is-feminine dualism, since these feminine attributes have been previously defined by men (Rose, 1993).

Whilst these writers would define themselves as ecological feminists (perhaps with the exception of Rose), and are writing from the position

Figure 3.1 *I'm your mother, not your maid!*
Source: Ontario Advisory Council on Women's Issues: Women and the Environment.

of a patriarchal critique, they question some of social and cultural eco-feminism's claims such as essentialism and the fact that the relationship between women and nature is not yet satisfactorily explained (Plumwood, 1993: 21). Other writers are more challenging in their criticism. For example, Cecile Jackson, researching environment and development issues, criticises eco-feminism for idealising women and for obscuring the differences and conflicts between them. She particularly identifies the tension between women in the North and South, where the former tend to construct the latter as 'victims'. She also, however, argues that in the South, women's relationship to each other and to the environment is heavily influenced by their class, age and position in the family and that any of these factors might result in treating the environment less favourably than eco-feminists would expect (Jackson, 1993).

Janet Biehl is one of the most vocal critics of eco-feminism and, as a self-proclaimed former eco-feminist (she writes from a social ecology perspective),[2] her criticisms are debated widely in this literature. Whilst she agrees with many of eco-feminism's arguments concerning the undervalued roles that women play in society, the hierarchies men have built on this and the lack of a sense of Western culture's dependence on nature, there are aspects of eco-feminism which have caused her to publicly break away from this position.

Primarily, she argues that certain aspects of Western culture, such as emancipation, democracy, civil society and public space, are positive legacies and should not be discarded, as she fears some eco-feminists are suggesting (Biehl, 1991: 1 and 136). She is also deeply suspicious of espousing goddess worship arguing, through a study of neolithic and Egyptian culture, that goddesses can still be hierarchical (ibid.: 37–48). She believes that such an emphasis is unrealistically backward-looking and urges us to look forward, taking the best of current society and changing the hierarchical, dominant worst. Finally, Biehl disagrees that women are special custodians of nature (ibid.: 5) and she looks forward to a society in which both women and men aspire to an 'ethic of humanity' (ibid.: 156).

Although Biehl has disassociated herself from the eco-feminist analysis, a number of other writers who share some of her reservations believe that the philosophy can incorporate these challenges. Plumwood and Warren both recognise that ecological feminism is not a unitary position and that it encompasses a range of opinions. Carolyn Merchant offers her own fourfold typology of eco-feminisms as a reformulation of the more diverse summary by Hallen, both of which can be seen in Boxes 3.2 and 3.3. These can be broadly placed along the social–cultural continuum described earlier in this chapter. Finally, Warren (1994) argues that ecological feminism needs to be inclusive and pluralistic, giving space to many voices, particularly the oppressed.

Summary

Whilst ecological feminism is not a widely known or understood philosophy, it is becoming increasingly important in critiques of human–society–nature relations more generally and is finding its way into a number of more general texts (e.g. Pepper, 1996; Simmons, 1997).

This chapter has presented eco-feminism as a critique of exploitative relationships between men and women, society and nature. Specifically,

Box 3.3

Feminism and the environment

	Nature	Human nature	Feminist critique of environmentalism	Image of a feminist environmentalism
Liberal feminism	Atoms Mind/body dualism Domination of nature.	Rational agents Individualism Maximisation of self-interest.	'Man and his environment' leaves out women.	Women in natural resources and environmental sciences.
Marxist feminism	Transformation of nature by science and technology for human use. Domination of nature as a means to human freedom. Nature is material basis of life: food, clothing, shelter, energy.	Creation of human nature through mode of production, praxis. Historically specific not fixed. Species nature of humans.	Critique of capitalist control of resources and accumulation of goods and profits.	Socialist society will use resources for good of all men and women. Resources will be controlled by workers. Environmental pollution could be minimal since no surpluses would be produced. Environmental research by men and women.

Cultural feminism	Nature is spiritual and personal; conventional science and technology problematic because of their emphasis on domination.	Biology is basic; humans are sexual reproducing bodies; sexed by biology/gendered by society.	Unaware of inter-connectedness of male domination of nature and women. Male environmentalism retains hierarchy. Insufficient attention to environmental threats to women's reproduction (e.g. chemicals, nuclear war).	Women/nature both valorized and celebrated; reproductive freedom; against pornographic depictions of both women and nature. Cultural eco-feminism.
Socialist	Nature is material basis of life: food, clothing, shelter, energy; nature is socially and historically constructed; trans-formations of nature by production and reproduction.	Human nature created through biology and praxis (sex, race, class, age); historically specific and socially constructed.	Leaves out nature as active and responsive; leaves out women's role in reproduction and reproduction as a category. Systems approach is mechanistic and not dialectical.	Both nature and human production are active; centrality of biological and social reproduction; dialectic between production and reproduction. Multi-leveled structural analysis. Dialectical (not mechanical) systems. Socialist eco-feminism.

Source: Merchant, 1996: 6.

eco-feminism challenges the dualistic philosophy of Western thought, particularly patriarchy. The chapter has illustrated the two broad perspectives of cultural and social eco-feminism but it must be stressed that these should be viewed more as ends of a spectrum and that these perspectives are changing as the philosophy itself develops. Whilst there are critiques of eco-feminist postures from both inside and outside the movement, this does not discredit the philosophy as a valid mode of inquiry. As Salleh suggests, by the late 1980s, eco-feminism was expressing a significant challenge to the 'transnational structure of capitalist oppression' (Salleh, 1995: 21). Much of the appeal of eco-feminism lies with it not only being a critique, but a normative philosophy (that is, arguing what should be rather than simply explaining what currently is), in that it looks towards a more harmonious relationship between society and non-human nature, however routes to achieve this may differ.

Notes

1 Menarche is the period which covers the beginning of menstruation in young women.

2 Social ecology is a movement in which ecology and anarchism are brought together in both a process and a set of goals which aim to end all forms of domination (between humans and between human and non-human nature). Social ecology advocates that we should live in mostly self-contained, small, ecologically self-reliant and confederal communities in which decisions are made collectively. In these, human impact on the environment is minimised. Murray Bookchin is the best known social ecologist. See Chapter 2 in Doyle and McEachern's book in this series (1998).

Discussion question

1 Which of the eco-feminist approaches appeals most to you, and why?

Further reading

Biehl, J. (1991) *Rethinking Eco-feminist Politics*, Boston, MA: South End Press. Although this book is quite hard to get hold of now, it is excellent if you would like to follow up her critique of eco-feminist philosophy.

Merchant, C. (1996) *Earthcare – Women and the Environment*, London: Routledge. Merchant is one of the best and most well-known writers in this area and the chapters in this book summarise her work over the last twenty years: including her first book *The Death of Nature: Women, Ecology and the Scientific Revolution* which examines how Western society has systematically relegated women in its search for control over nature. She also offers a useful discussion of different kinds of eco-feminism which first appeared in her book *Radical Ecology*. In conclusion, Merchant makes a powerful case for what she calls 'Partnership Ethics: Earthcare for a new millennium'.

Plumwood, V. (1993) *Feminism and the Mastery of Nature*, London: Routledge. This is a useful book to read to follow up the nature of dualistic thought and its historical background. Plumwood argues for the transcendence of dualistic thought.

The body and the environment

Key words: sex/sexual norms; medicalisation of the body; food entitlement/distribution; gendered mortality rates; chemical pollution

- **Normal male/deviant female**
- **Wise woman to medical men**
- **Missing women**
- **The gendered effects of pollution**

It is through our bodies that we most directly confront our environment. Carolyn Merchant has even obliterated the distinction between self and environment, calling the body the 'primary environment' (Merchant, 1996: 223). Before examining how, in this 'primary environment' we are affected by the secondary environment depending on our maleness or femaleness, it is important first to examine what exactly is meant by the body and how its male and femaleness is constructed. After this construction has been explored, the chapter will look at the medicalisation of gender and the gendering of medicine, particularly 'female problems'. At the most basic level, even survival is affected by sex and the differences in male/female mortality and well-being will be considered. The health of our bodies is also dependent on a chemically safe environment. Where pollution levels are unsafely exceeded we may suffer discomfort, disease and even death. Different kinds of pollutants have various effects on different bodies: male and female, child and adult, as the examples of oestrogen pollution which affects sex and fertility and the leakage of methyl isocyanate from the Union Carbide plant in Bhopal, India, shown later in this chapter will graphically illustrate.

The uses to which we put our bodies also opens us to different kinds of environmental threats, whether this is coalmining, working in a chemical factory, prostitution or collecting water and firewood. Chapter 5 will explore how our social and economic roles expose us to environmental damage and how this is gendered.

Normal male/deviant female

Chapter 2 showed how Aristotle's biological theory of male superiority viewed the 'female of the species as an incomplete and mutilated male' (Merchant, 1983: 13). His views find echo in the example given in Chapter 2 of how maleness in nature (sperm) is seen as positive and dynamic whilst femaleness is seen as partial and degenerative. Aristotelian belief was that the passive female reproductive capacity was activated by the dynamic semen and this has gained currency since the seventeenth century as medical interventions became increasingly commonplace. Moreover, menstrual blood was considered unevolved semen, lacking the necessary 'warmth' to be transformed. Whereas these concepts are no longer held to be valid, Chapter 2 also illustrates the 'otherness' of women, whereby women's reproductive mechanisms are considered by the sub-discipline of gynaecology and women's reproductive and ageing processes are frequently construed as problems which need fixing by modern medicine, as the following section will develop.

As Chapter 3 concluded, there is considerable debate over the degree to which biology determines socialisation and we should also take care in describing bodies as either male or female, pausing to consider the ambiguities of the sexed body. Judith Butler interrogates the fixedness of sex, arguing that not only gender is mobile (see Chapter 1) but also sex. She questions how 'sex' is defined: is it anatomically, chromosomally, or hormonally? By asking how the definition of sex arose in the first place, what might at first be understood as a 'natural fact' can also be seen as a product of political and social interests (Butler, 1990). If this point is accepted – that sex as well as gender is socially constructed, then any attempt to ascribe sex-specific attributes (as in the case of essentialist/cultural eco-feminism) becomes impossible.

It is not always as easy to recognise the 'flexibility of the continuum along which sexual differentiation occurs' (Epstein in Cream, 1995: 34), and Julia Cream attempts to clarify the grey area between male and female as containing a number of ambiguous states such as 'hermaphrodites', 'androgynes' and transsexuals. An autobiography of Lady Colin Campbell, published as this book was being prepared (1997), tells of a masculine childhood chosen for her by her parents as she was born without a vagina (the doctors deemed that any ambiguously sexed child be classified male – another example of the male 'norm'). Subsequently, as a teenager she underwent operations to create external

Box 4.1

The third sex

When a baby is not a boy. Nor a girl.
By Jay Rayner

Newborn babies are experts at keeping secrets, their fragile, creased bodies and clenched eyes conspiring to tell us next to nothing about the selves they are to become. The only substantial clue any new parent can hope for lies with the brutal segregation of gender. A quick check in the right place and the cry goes up: congratulations, it's a boy; congratulations, it's a girl. This much you are allowed to know.

Some parents, however, are robbed even of this fragment of knowledge. A rare hormonal imbalance during pregnancy can affect the physical development of the genitals so that their shape becomes indeterminate. The child possesses a tiny penis, say, or an enlarged clitoris, and in these circumstances it is fiendishly difficult to be certain to which gender the new arrival belongs. The condition is called intersex, and affects one child in 12,000, or about 60 births in Britain each year.

'This really is the most awful thing that can happen to anybody,' says Professor Charles Brook, who heads the intersex clinic at Great Ormond Street hospital in London, the only one of its kind in Britain. 'It's a question that will affect people for life, and one to which, frankly, we do not have all the answers.' The condition, which can be caused by a malfunction of the mother's adrenal gland, can result in babies who are either chromosomally female but masculinised by exposure to male sex hormones or, conversely, chromosomally male but under-masculinised.

At Great Ormond Street there is a team of surgeons, gynaecologists, urologists, endocrinologists and psychologists who together try to decide, within the first year of a child's life, which gender it should be placed in. They use a combination of therapies, including reconstructive surgery of the genitalia and hormone treatments. 'Genetic females are generally raised as females after surgery,' says psychologist Melissa Hines, who works with Brook's team, 'because they usually have internal female organs.' However, it's not always that simple.

The hormonal imbalance which leads to the condition can affect both brain and behavioural development as well as physical development. In short, the child can end up feeling male, despite being placed within the opposite gender. The physicians are therefore required to make the right call very early on in a child's life, when little about its future identity has been established, and when there are very few dues upon which to draw.

'It is a difficult, sweaty area of medicine to be in,' Brook says. 'And it's not until you're quite elderly that you see children coming to maturity and you can tell whether you have made mistakes or not.' Intersex cannot therefore be thought of as an issue to be dealt with solely at birth. The children have to be helped to understand their own condition; they may need hormone injections to take them through puberty; and some – though certainly not all – may find themselves sterile. Brook admits that he knows of mistakes being made but, he adds, many intersex people can go on to lead perfectly normal lives.

In the US, there is currently a debate over whether medical intervention to assign gender to intersex children is justified. A number of academics argue that the condition is evidence of more than just two rigid genders, and that a physical blurring of the distinction is something for society to deal with, not the individual. 'The position in America is all to do with knocking doctors off their pedestals,' says Brook, 'and it is something with which I have a certain sympathy. As I see it, if you had enough peers in these third or fourth genders, then it would seem a reasonable approach to leave the children as they are. Otherwise, it would strike me as being a rather lonely existence.'

Or, as Melissa Hines puts it: 'If we lived in a social vacuum, non-intervention would be best, but society expects us to conform.' It seems that in certain fields medicine becomes less a science than an art and, like all great arts, it can only ever aspire to perfection. Intersex, with its unique social and philosophical problems, is one of those fields of medicine.

Source: Courtesy of the *Observer*.

female genitalia, but she remains unable to conceive. Box 4.1 reproduces an extract from the *Observer Life* which reports on a condition known as intersex, affecting one in twelve thousand British births. Whilst medical experts attempt to assign a sex to the intersex child as soon as possible, there is currently a debate in the USA which argues that intersexuality demonstrates the existence of more than two sexes (see Box 4.1). However, as examples of how pollution affects the body's physiology will show later in this chapter, the existance of a sex continuum is, perhaps, more appropriate than the two polar opposites of male and female.

Wise women to medical men

Early medicine was largely the preserve of women whether it be through the adminstration of herbal remedies or midwifery. Up until the seventeenth century, it was considered 'improper' for men to be present at childbirth . However, the development of forceps in the seventeenth century required training and licensing and since this was only available in medical schools, which did not admit women, women became increasingly excluded from attending childbirth. By the end of the century, male surgeons dominated the practice and the body was increasingly being seen as a 'machine' which required tuning and correction by medical intervention (Merchant, 1983: 151–156).

By the later nineteenth century, the field of gynaecology had been established as a separate branch of medicine to study the 'natural woman' (that is, specifically female processes). As Oudshoorn has pointed out, this reinforces the notion that women are aberrant males as no equivalent branch of medicine was established to study specifically male processes (Oudshoorn, 1996: 157). Since then, female 'processes' have increasingly come under medical scrutiny with a tendency to be coined as 'problems'. Thus reproduction is a female 'problem' to be controlled by the development of sterilisation, interuterine devices or 'the pill'. Menopause is a 'problem' to be disguised by the use of hormone replacement therapy and menstruation is a 'problem' to be dealt with by developing a super-absorbant tampon which, as well as making menstruation invisible, can cause 'toxic shock syndrome'. (See Garrett, 1995 and Figure 4.1, which also emphasises the environmental costs of sanitary protection.)

Ironically, it is the entrance of women into the public spaces of paid work and the fight for equal opportunities there which has encouraged these technologies. The women they are supposed to benefit do so, however, at the expense of less advantaged women. Oudshoorn describes the clinical trials of the contraceptive pill in the 1960s. Designed to be a universally applicable medication, the pill was developed in US laboratories and yet all the major trials were conducted in the Caribbean (three in Puerto Rico, one in Haiti). Consequently the risks were born by a relatively powerless group of women whose preferred method of birth control was, and still is according to Oudshoorn, sterilisation and the interuterine device (Oudshoorn, 1996). Mies and Shiva go further in their criticism, arguing that all such methods of contraception have, to some degree, been forced on women in the South. This coercian has, they say, been undertaken to curb 'the problem' of 'over-population'. Mies and Shiva discuss the way in which institutions such as the United Nations and the World Bank conceptualise fertility as a 'disease', or even 'epidemic', which Western governments resolve to control through administering

Figure 4.1 *With a tampon, every day's fun*
Source: Courtesy of the Women's Environmental Network.

contraception programmes. In support of their argument, Mies and Shiva cite the sterilisation programmes in India and Bangladesh in which sterilisation is exchanged for food, suggesting that the Vulnerable Group Feeding Programme has been used to force the poorest women to be sterilised (Mies and Shiva, 1993: 191).[1] Arguments which blame environmental degradation on over-population are often held in contempt by Southern commentators who protest that this places the blame for environmental problems on societies with often very low per capita resource use. There is also a counter-argument to the assertion that over-population is the main cause of environmental degradation.

It is not only women in the South who are coerced into receiving medical contraception. In 1989, the UK-based WHRRIC discovered that Depo-Provera (a contraceptive administered by injection) was routinely being given to black and Asian women in Leeds (Lovenduski and Randall, 1991: 230–231).

In their discussion about reproductive technologies, Mies and Shiva argue that women's bodies have been reduced to a set of 'fragmented, fetishized and replaceable parts, to be managed by professional experts' (1993: 26). This management includes not only fertility control, but techniques such as IVF (in vitro fertilisation) and artificial insemination. Thus 'reproductive technology alienates both men and women from their own bodies and from the most intimate process in which they normally cooperate with their own nature' (ibid.: 139).

On the one hand, traditional, 'natural', methods of contraception such as 'coitus interruptus' or the use of medicinal herbs, are replaced by an industrialised technology whilst, on the other, men and women's fertility is often inhibited by environmental factors. This will be examined later in this chapter, after the differential survival rate of women and men is considered. As well as being under pressure to reduce the number of children they give birth to, women have, at times, been exhorted to devote their bodies to the nationalist cause. Nationalist propaganda has, for example, encouraged Israeli women to have children to increase the Jewish birth rate to compete with that of the Palestinians, whilst in Australia white women were encouraged after the Second World War to be 'walking wombs' by one politician (Cockburn, 1998; Pettman, 1996).

Missing women

Everyone needs a basic intake of air, food and water to survive, so if these are contaminated or unavailable our health and even survival are threatened. Even at this basic level a person's sex may determine their survival and well-being. Jean Dreze and Amartja Sen's work on famine illustrates how food within the family is allocated both during 'normal' and 'temporary distress' situations. They reveal widespread female nutritional disadvantage, particularly in South and West Asia, North Africa and China. Whilst women seem to have a greater capacity than men to cope with temporary distress situations (such as famine) because of their biological characteristics, they are more likely to suffer relative deprivation in normal situations. Food priorities in distress situations are found to be pro-male and whilst women suffer a lower mortality rate than men, their physical distress is greater. For example, Fernandes and Menon's research in Orissa, India, quoted by Dreze and Sen, states that 'during scarcity, children get first priority, then come men and then only women', a bias also found by Médecins sans Frontières in the 1983–1985 Sahel crisis (Dreze and Sen, 1989: 79).

Two groups are identified as suffering disproportionate food deprivation in the early stages of subsistence crisis: the elderly and adult women. The latter bear a disproportionate share of the burden of adjustment in comparison with adult men, although they do not typically experience a higher increase in mortality. Whilst young children appear to be comparatively protected in Sub-Saharan Africa, with no boy/girl distinction, China and India both demonstrate greater protection of boy children (Dreze and Sen, 1989: 81). Nor are food shortages for women confined to the South. Research on the UK reports that half of all mothers living on, or just above, the income support level regularly went without food to feed and clothe their children (Kempson *et al.*, 1994).

Around 105–106 boys are born for every 100 girls around the world (Bandarage, 1997: 97). In addition, fewer females survive into adulthood than males in most of the developing world, with mortality rates for girls higher than those for boys (see Table 4.1). This is due to a number of reasons including likely discrimination against girl children with regard to food and health care.

Table 4.1 *Female/male ratios, infant and child mortality rates, selected countries, 1986/1993*

Region	Female/male ratio	
	Infant mortality	Child mortality
Developed regions (30 countries)	0.8	0.8
Developing regions (Northern Africa and Western Asia)		
Egypt	0.9	1.4
Jordan, Morocco, Tunisia, Yemen	0.9	1.1
Sub-Saharan Africa (17 countries)	0.8	1.0
Latin America and Caribbean	0.8	1.0
Asia		
China	1.2	1.0
Pakistan	0.8	1.6
Indonesia, Philippines, Sri Lanka, Thailand	0.8	1.0

Source: United Nations, The World's Women, 1995: Trends and Statistics, Social Statistics and Indicators, series K, no. 12, New York, 1995: 69; in Bandarage, 1997: 173.

The higher value put on males discriminates against nuturing girl children where they may be in competition for resources with their brothers. Bandarage cites Caldwell and Caldwell's research in South Asia as identifying that 'females receive less care, fewer warm clothes, less medical attention (and that given belatedly) and in spite of the drain of pregnancy and lactation, less food' (Caldwell and Caldwell, 1990: 17, in Bandarage, 1997: 173). Table 4.2 demonstrates the statistical result of such gender discrimination, showing the number of 'missing women' in different parts of the world.

More sinister than this is the deliberate choice to abort a female foetus and the killing of baby girls. Maria Mies and Vandana Shiva are particularly concerned about the ability of modern reproduction and fertility techniques (such as amniocentisis) to identify the sex of the foetus. One outcome of this is the abortion of female foetuses in cultures where girl children are seen as second best (Mies and Shiva, 1993:194). Female infanticide has a long tradition in patriarchal societies such as India and China, and recent Chinese family planning policies have reinforced this. Abortion after sex determination is thought to be still widespread in China and India, although, technically, the use of sex determination is illegal in China for the purpose of deciding whether or not to abort a female foetus. Induced abortions were recorded at 53 per cent of all births in 1986, rising from 31 per cent of all births in 1978 (Bandarage, 1997: 99).

Table 4.2 Missing women by region

Region	Female/male ratio	'Missing women' in millions	%
Europe	1.050		
Sub-Saharan Africa	1.020	Taken as the 'norm'	
SE Asia	1.010	2.4	1.2
Latin America	1.000	4.4	2.2
North Africa	0.984	2.4	3.9
West Asia	0.948	4.3	7.8
China	0.941	44.0	8.6
India	0.933	36.9	9.5
Pakistan	0.905	5.2	12.9

Source: Dreze and Sen (1989: 52).

Although sex discrimination tests are now outlawed in several Indian states, it is a thriving business elsewhere with 78, 000 female foetuses recorded as being aborted after sex determination tests between 1978 and 1983 (Bandarage, 1997: 99). Abandonment of girl children is also not uncommon: most of the children who end up in Chinese state orphanages are girls and once there, they are likely to be abused and neglected. Bandarage quotes a Human Rights Watch report which claims that over 1, 000 children died between 1986 and 1992 due to the brutal treatment at a Shanghai orphanage (Bandarage, 1997: 100).

Violence against females is not, of course, confined to girl children. UNICEF claims in 'The State of the World's Children' that violence against women by male partners is the most common crime in the world (UNICEF in Pettman, 1996: 185). Moreover, in war, women are increasingly likely to be injured or killed. Ninety per cent of war casualties are currently civilian, the majority of whom are likely to be women and children. This has been steadily increasing: in the First World War, this figure was 1 per cent, in the Vietnam War it was 80 per cent. Assaults against women's bodies are also used as a military strategy. In the Yugoslav civil war up until 1995, variable estimates of 20, 000 to 35,000 rapes were recorded (Pettman, 1996: 94). Rape was never considered a war crime until 1993 when the United Nations adopted the Declaration on the Elimination of Violence against Women. This declaration defined what constituted violence against women and outlines actions which governments should take to prevent such acts.

These are *proximate*, or immediate, causes of 'missing women' in the Third World, which can, in turn, be attributed to *fundamental* causes which result from a society (which feminists describe as patriarchal) in which men are valued more highly than women. Women of all classes are under pressure to produce sons. Most decisions on how many children to have are made by a women's husband and are reinforced by the extended family. Women's worth is measured by their fertility, and anticipating this they are given an exchange value in the form of a dowry. In extreme cases, 'dowry death' is not unknown (where a bride is killed because the wealth transfer from the bride's family to the husband is thought to be inadequate). In India, 1, 786 such deaths were reported for 1987 (United Nations, in Bandarage, 1997: 172). Particularly in the Third World, girls' and women's bodies are vulnerable in a society which places a higher premium on the male. This can be seen in the way in which females are seen as a departure from a male 'norm' and as the body responsible for human reproduction. Their survival rates are not only less than men's but are a direct result of a society which prioritises men. The next section moves on, in the light of this, to consider the way in which environmental problems affect the bodies of males and females.

The gendered effects of pollution

As the eco-feminist arguments in Chapter 3 have asserted, women tend to identify more closely with their bodies than do men, although there is a debate as to whether this is due to biological features such as menstruation, child-bearing and nursing (cultural eco-feminism) or to women's social roles as the main provider of food, water, health care and physical comfort (social eco-feminism).

Chapter 5 will consider the gendering of environmental effects with regard to social roles. Regardless of role, however, it is clear to see how pollution has a distinct effect on the 'primary environment' of the sexed body. Men's and women's bodies respond differently to environmental pollutants, either because of the nature of the chemicals and their variable interaction with physiological differences or because of physical vulnerabilities linked to issues covered in the previous sections. For example, in the Love Canal environmental protest against a toxic waste dump (see Chapter 6 for more detail) inadequate guidelines on the effects of chemical pollutants were given for women and children as the only available estimates were based on the assumption of workplace exposure of forty hours a week on men, who have a heavier body weight (Gibbs, 1998). Different occupations can also lead to gendered effects of pollution. For example, US servicemen who had been exposed to Agent Orange (the powerful defoliant used to reveal enemy forces below the rainforest canopy) in Vietnam, suffered from dioxin poisoning (not to mention the Vietnamese themselves). Dioxin, 'one of the deadliest poisons known to man' and 'one of the most potent carginogens that had ever been tested' (Cadbury, 1997: 190), had contaminated one of the component chemicals of Agent Orange. It also affects male fertility and causes chloracne, a debilitating skin disease.

The persistent use of chemical pesticides incorporating aldrin, benzene, chlordane, dieldrin, heptachlor and (where not banned) DDT, causes female infertility and contaminated breast milk. Merchant reports a high percentage of miscarriages and birth defects in New South Wales, Australia, where some of these pesticides were aerially sprayed on crops (Merchant, 1996: 194). DDT is a widespread contaminant of human milk, first brought to popular notice by Rachel Carson in her book *Silent Spring* (1962) Although initially vilified by the agro-chemical industry (for example, Monsanto accused her of being an 'emotional female alarmist'), Carson's work was instrumental in effecting a ban on DDT in the West (Poklewski Koziell, 1999). Because it is stored in fat tissue for years, DDT is passed to offspring through the placenta before birth, and afterwards through the mother's milk. In areas where DDT is now banned, levels of 30ppb (parts per billion) have been measured, but levels rise up to one hundredfold in areas where this pesticide is still used. For example, in Guatamala in 1970, levels of 100ppm (parts per million) were recorded (Cadbury, 1997: 172). Children who are passed this chemical load are at increased risk of a number of diseases including endemetriosis and breast cancer (in female offspring) and reduced fertility. Sascha Gabizon reports from a visit to the Aral Sea region of the former Soviet Union, how high concentrations of pesticides (and heavy metals) have been found in what remains of this once huge inland sea. Defoliants used in cotton production are thought to be causing blue baby syndrome and mothers have been advised to switch to bottle feeding. In the adjoining republic of Karakalpakstan the highest levels of maternal and infant mortality in the former Soviet Union were recorded and these figures rose with proximity to the Aral Sea. An environmental justice issue raised here is the anaemia of breast-fed babies (where mothers have high levels of DDT and lindane recorded) which is exacerbated by poor diet and hygiene (Gabizon, 1998: 8–9). The environmental justice argument is also important in understanding why some populations are exposed to dangerous chemicals whilst others are not. For example, whilst DDT has been banned in the West since the 1970s, this has not been the case elsewhere.

DDT and lindane are members of a group of chemicals identified as hormone disruptors (that is, they interfere with key hormones which affect sexual development and fertility: oestrogen, progesterone and testosterone) which have a profound and gendered effect on health. Research in the United Kingdom indicates that oestrogen is emanating from a number of different sources. Natural oestrogen finds its way into the water supply through the sewage system (female excreta), but this is added to by the use of ethynylestradiol used in the production of oral contraceptives, which are higher in concentration and longer living than naturally produced oestrogen. Oestrogen mimicking chemicals are also found in polychlorinated biphenyls (PCBs), surfactants (used in cleaning products) and, one of the most common artificial chemicals in the environment, phlalates, which are used in the manufacture of plastics to enhance flexibility (such as PVC and cling film). From all these sources artificially produced oestrogen can leach directly into water, soil and food, or indirectly through landfilling. Organo-chlorine pesticides such as dieldrine are also thought to have an oestrogenic effect on mammals. One effect of these chemicals has been identified as a major drop in fertility. Whilst this has yet to be proved to the satisfaction of legislators and chemical products manufacturers, circumstantial evidence presented by Danish research suggests that this may be linked to a 50 per cent drop in sperm count recorded over the last fifty years, a proportion upheld in studies elsewhere, whilst testicular cancer, another associated illness, now affects more than 1 per cent of all men in Denmark (Cadbury, 1997: 206).

Research undertaken on fish suggests that, in addition to causing reproductive disorders, the rise in circulating oestrogenic chemicals can affect the sex of a fish. Male fish exposed to oestrogenic substances manufacture a yolk protein (vitellogenin), not normally found in males, but almost identical to that naturally found in mature females (Harries *et al.*, 1997). Research by the same team has found an extremely high percentage of fish to be hermaphrodite (Sumpter and Jobling, 1995). Cadbury reports on research which has identified a fungicide (vinclozolin) as reducing testosterone in males and, in experiments, rats have become biologically 'feminised' when exposed to miniscule quantities (Cadbury, 1997: 187).

The main problem with these chemicals is that they have a cumulative effect in the body. Whereas naturally produced oestrogen and oestrogen taken in the contraceptive pill stays in the body for a matter of hours, oestrogen-mimicking chemicals absorbed through eating fruit and vegetables treated with pesticides, or stored in tins lined with bisphenol A or through amalgam used in white fillings, remain in the body for years. Scientists interviewed for a Horizon TV documentary by Deborah Cadbury believe that there is sufficient accumulated evidence (though not proof) that these chemicals are responsible for a large proportion of breast cancer, 90 per cent of which is thought to be caused by environmental/lifestyle rather than genetic factors.

One in twelve women in the UK is currently diagnosed with breast cancer and in some areas of the USA the figure is as high as one in eight. This compares with one in twenty-two fifty years ago (although some of this rise will be accounted for by better monitoring). Studies in New York have established that, in their study sample, women with the highest 10 per cent of exposures to DDT had a fourfold increased risk of breast cancer (Cadbury, 1997: 229).

All this has a potentially significant and differential effect not only on the health of men and women, but on sexual characteristics. Though more research needs to be done in this area, there is enough circumstantial evidence to suggest a link between rising reproductive disorders and abnormalities and increases in the amount of oestrogenic substances (Harries *et al.*, 1997). This does, however, question the immutability of sex, which was referred to earlier. The fact that environmental factors can change the biology of the body from what society defines as male to that which is defined as female raises the interesting point of how far we can fix male and female as opposing categories.

As well as the routine contamination of our bodies, environmental disasters can also have a differential effect on men and women. The crisis in Bhopal, India, where, in 1984, a Union Carbide pesticide plant leaked methyl isocyanate gas, chloroform and, as some observers still argue, hydrogen cyanide, shows how the physical effects of this had a clear, body-based, gender dimension. An estimated 3, 000 people died within a week of this leak and a further 30, 000 were estimated to have been affected by exposure to the poisonous gases. It is also clear that poverty had a significant impact on the death and illness toll.

Most deaths were caused because people either fled their homes or were living in extremely flimsy structures which offered no protection against the gases. (Twenty per cent of Bhopal's population lived in slums and shanty towns, two of which were directly across the street from the Union Carbide plant.) Those who died were amongst the poorest people in the city. Those who survived suffered respiratory problems (including bronchitis, pneumonia, asthma and fibrosis), digestive disorders, severe eye irritations and, amongst women, menstrual disorders and the suppression of lactation in nursing mothers. Women were most likely to suffer psychological problems, which were most severe in women of child-bearing age. High infant mortality/morbidity rates exacerbated this. Because a high regard socially is held for women's fertility, these psychological problems were said to be partly caused by women not wanting to admit gynaecological illnesses to their family for fear they would be devalued as a bride, or unmarriageable (Shrivastava, 1992).

In addition to ill health related to bodily function, women's roles also exposed them to additional environmental damage. On average a women would spend four to six hours a day preparing food on wood or coal-burning stoves in poorly ventiliated, smoke-filled kitchens. This continued inhalation of smoke worsened the damage the Union Carbide leak caused to women's lungs and eyes.

Poverty has also been found to have had a compounding effect on the health damage by radiation from the explosion at the Chernobyl nuclear power plant in the Ukraine. Marie Kranendonk reports how these might well be worsened by 'synergetic effects of an overall high chemical pollution and a continuous increase in poverty among the general population' (Kranendonk, 1998: 6–7). The links between poverty and gender will be explored in Chapter 6 when women's economic position in relation to the environment is considered.

Summary

This chapter argues that there are both proximate and fundamental causes for the ways in which our bodies experience the environment. In terms of considering why fewer girls than boys survive birth and childhood a number of proximate causes were given such as unequal food distribution, abortion of female foetuses and female infanticide. Fundamental causes include those patriarchal social structures which value men over women and boys over girls. In terms of chemical pollution, diseases can arguably be traced to the increased use of pesticides, fungicides, herbicides, plastics and so on. However, the fundamental cause, again, is a social structure which is dominated by industrial interests and capital. This inequality of the effects of environmental degradation has given rise to a substantial literature on environmental justice which examines the differential effect of environmental degradation on rich and poor, black and white, and it is within this context that gendered differences of environmental impacts should also be viewed. (For a good summary of this literature see Dobson, 1999.)

Scientists researching these issues, even though they are mindful of jeopardising industrial cooperation and funding, are challenged by an industry which is an extremely powerful force in the global economy. Sales of synthetic chemicals and products derived from them constitute well in excess of one-third of the world's GNP (Cadbury, 1997: 255).

Finally, the chapter has indicated that the body itself is not an uncontested concept and that we need to be wary of simple assumptions as to what defines a male or female body. What constitutes 'normality' in a body is heavily influenced by the prevailing norms in society, although we should be wary not to allow this challenge to a male/female dualism to 'normalise' the sexually distorting effects of certain chemical pollutants. Scientists are unlikely to be preoccupied with these debates and we need also to be aware of the parameters within which science works, as discussed in Chapter 2. Just as 'the body' should not be regarded without analysing what is meant by this, the same is true of the positions people assume or are ascribed to in society. The next chapter looks at how environmental issues impact on different family or household members depending on their position within it.

Discussion questions

1 Reflecting on your own experience and/or that of people close to you, consider how female bodily processes have been construed as 'problems' by the medical profession and the drugs industry.

2 Consider all the ways in which boy children are prioritised over girl children. How is this likely to perpetuate unequal gender relations?

3 With reference to forms of environmental pollution not covered in this chapter (e.g. CO pollution from cars), examine how female bodies may be affected differently from male bodies. You might also assess the effect on children's bodies or the bodies of elderly people.

Further reading

Bandarage, A. (1997) *Women, Population and Global Crisis*, London: Zed Books. This is a good account of how women, particularly in the South, are affected by mainly global processes emanating from Western countries, institutions and companies.

Cadbury, D. (1997) *The Feminization of Nature: Our Future at Risk*, Harmondsworth: Hamish Hamilton. Deborah Cadbury is a scientific journalist and TV producer who has written this very readable account of how scientists are uncovering increasing evidence of adverse changes in human reproduction and health. This was also the subject of a Horizon TV programme in the UK which she produced in 1993.

Lykke, N. and Braidotti, R. (1996) *Between Monsters, Goddesses and Cyborgs: Feminist Encounters with Science, Medicine and Cyberspace*, London: Zed Books. This book has already been recommended in Chapter 2, but Part Two,

'Biomedical Bodygames', is especially interesting with regard to the medicalisation of particularly female body functions.

Mies, M. and Shiva, V. (1993) *Ecofeminism*, London: Zed Books. This is an interesting book co-written by a sociologist (Maria Mies) and a physicist (Vandana Shiva). Part 4, 'Ecofeminism v. New Areas of Investment through Biotechnology', is particularly relevant to the material presented in this chapter. In this the authors discuss the implications of new reproductive technologies to women in the North and the South.

Note

1 It is, however, interesting to note that when women become more financially independent, their fertility appears to fall, as Chapter 7 will show in the context of micro-credit schemes designed to raise the independent income of women in poverty. What I am discussing here, however, is the coerced administration of birth control.

5 ▶ Gendered roles in the family

Key words: household structure; domestic division of labour: public-private spheres; double-day; gendering of environmental impacts

- The organisation of the household
- The domestic division of labour
- The gendering of environmental impacts by household roles

Chapter 2 explained how our beliefs, understanding and practices are shaped by prevailing social, political and moral frameworks. One powerful unit which participates in this shaping is the family. For the young child, the household (group of people living together, not necessarily related by blood) or family (group of people linked by blood ties) is the most important socialiser, although the size and shape of the family – household changes over time and through space. In this chapter I will first look at the changing nature of the family – household and the roles of women and men within this before turning to how these roles involve men and women in particular relationships with the environment.

The organisation of the household

The household has undergone several radical transformations through time. Alison Hayford, in an early contribution to *Antipode* (the radical geography journal launched in 1968), wrote about the transformation of the earliest household group into a unit which supported capitalist production in the nineteenth-century West.

> The household is the main means that human groups which are active on a relatively large scale of production have devised to lessen the tensions of space. . . . Whatever its form the household was, until the development of capital relations and the final dominance of the public sphere, central to

> the productive system of any group: from it people went out to work the
> earth, to it they brought what they cultivated or made, within it and from
> it they distributed these things. Since women were at the core of the
> household[1] they were pivotal to all these functions, even where they had
> little formal political power . . . women remained at the centre of what
> was private.
>
> (Hayford, 1974: 138)

As you can read from this, women held a position of some centrality and
meaning in such households, not least because they represented linkages
between household groups. That is, in patrilineal societies (in which the
line passed down from father to son), a woman married into a man's
household and was expected to live there, bearing his children. This
contractual relationship represented a tie between two households,
although the woman's mobility was tightly controlled. (Hayford gives the
example of Chinese women who were entitled to visit their natal home
once a year and of Islamic women who may not leave their husband's
house for years, ibid.: 140.)

As states became politically formalised, the household was recognised as
an expression of state power. For example, in English law, a wife who
killed her husband was guilty of petty treason (as the husband was
considered the proxy head of state within the household), whilst in
China, the order of the Confucian system was maintained through the
order of the family (Hayford, 1974: 139).

With the development of a class-based society in the West, the control of
economic production was removed from the family and the ties between
households or families were replaced by the key relationship between the
household and the higher authority. Because of the redrawing of these
ties, women's role as the key link between households was diminished
and a significant role was lost. The Industrial Revolution in the UK and
later as it spread through Europe and North America resulted in all
economic production taking place in the public sphere outside the
household. Consequently what had previously been produced internally
to the household for use was increasingly produced externally for profit
with household members having to exchange their labour for the means
of survival through the factory system. In order for households–families
to survive it was frequently necessary for women (and sometimes
children) to also sell their labour, but even this failed to raise the standard
of living of many households, to the extent that by the mid nineteenth
century the household was breaking down as a social unit in industrial

countries. In Chapter 7, the increasing involvement of women (and children) in the paid labour forces of the South will show the now global reach of this form of economic organisation.

Reform movements of the late nineteenth and early twentieth centuries in the West were in part designed to re-establish the household as *hearth and home* (and as a means of political control). It is from this that emerged the widespread belief that women should not be involved in productive labour and the line between private and public space was made more explicit. Twentieth-century notions of the nuclear family containing a working father, home-based mother and a small number of dependent children are based on this ideal and it is important to recognise that the *traditional family*, to which some politicians would have us return, is a socially constructed response to conditions at this time. It is also important to remember that this *traditional family* is as much an ideal as a reality as a percentage of women have continued to work outside the home throughout the twentieth century. In 1950, the proportion of traditional families (that is, waged father, unwaged mother and one to four dependent children) in the USA was 60 per cent but by 1985 this had diminished to 7 per cent (Pratt and Hanson, 1993: 28). The atypical nature of the 'traditional family' is reinforced by Table 5.1 which shows that the majority of households do not have dependent children in them, and that just under a quarter of households with dependent children are headed by a lone parent.

Table 5.1 *Households by type of household and family in the UK*

Type of household	Percentage of all households
One person	27
Two + unrelated adults	2
Single family households:	
Couple, no children	28
Couple, 1–2 dependent children	21
Couple, 3+ dependent children	5
with non-dependent children only	6
Lone parent, dependent children	7
with non-dependent children only	3
Multi-family households	1

Source: HMSO, 1998.

The percentage of women in paid work has also differed depending on which group of women is considered. For example, in 1960 in the United States, 36.5 per cent of white women were employed outside the home compared to 48.2 per cent of black women (Smith, 1987 in Pratt and Hanson, 1991: 57), although by 1995 this gap had closed with 59 per cent of both black and white women classified as part of the civilian labour force (US Bureau of the Census, 1995).

Anthony Giddens argues that the family in the West is going through another transformation in the late twentieth century. Whereas the *traditional family* is being deconstructed because of the increase in lone parents, divorce, open homosexuality, and non-marrying, he suggests that there is potential for the *recombinant family*, negotiated rather than determined by blood ties, to be strengthened. In this family, inter-household ties are developed between step-families and between generations to support the raising of children and caring for vulnerable members. Giddens suggests that as marriage becomes emptied of its traditional meaning (that is, as an economic, contractual relationship), its survival rests on its capacity to nuture a mutually rewarding relationship (Giddens, 1994: 172).

Such potential, however, is not universally accepted and many feminists continue to see the home/household as a site of labour, albeit unpaid. Pratt and Hanson review the perspectives which argue that the household continues to support capitalism by requiring (mostly) women to provide unpaid labour which cheapens the costs of social reproduction (that is, absorbing costs which would otherwise have to be borne by capital or the state). This is true not only of the West, but more emphatically in the South.

So far, this discussion of the household and family has focused on its development and characteristics in the West. Elsewhere in the world, households are more likely to resemble the early economic unit that Alison Hayford portrays in the quotation at the beginning of this chapter. Cindi Katz, writing about households in rural Sudan, describes how the compound is an economic unit comprising parents, children, grandparents, aunts and uncles with their offspring. Katz does, however, point out that subsistence economies are now gradually being replaced by wage labour as development projects are introduced (Katz, 1993).

A looser kind of household is described by Lydia Pulsipher (1993) where she explores the 'yards' of the Eastern Caribbean. Here, anything up to thirty residents, not necessarily with blood ties and possibly including business partners, share a compound in which single-room dwellings are mixed with an outdoor kitchen, pens for animals, gardens and showers. Unusual, in that this is a matrilocal system, young mothers are not commonly joined by the fathers of their children, who remain at their mother's homes with responsibility for their nieces and nephews. Grandmothers and other members of the extended family will often care for children, enabling mothers to take paid employment. Such practice is reflected in the high percentage (70 per cent) of children born to non-married parents (Pulsipher, 1994: 114).

Whilst there are different forms of households in different parts of the world, all households are distinguished by the tendency for women to work a 'double day', that is to have responsibility for housework and care of dependants in addition to engaging in subsistence farming or in paid work. In the next section, I will explore the extent to which women continue to undertake unpaid domestic labour, but you should reflect on your own family and household to see the extent to which they fit these pictures.

The domestic division of labour

> *The world is, at present divided into two services: one the public and the other the private . . . Is the work of a mother, of a wife, of a daughter worth nothing to the nation in solid cash? . . . astonishing . . .*
> (Virginia Woolf, *Three Guineas*, 1992: 230)

These two spheres of the public and private have become common ways of delimiting the arenas of activity of men and women. Through their roles as mothers, carers and domestic workers, women are associated with the private sphere of the home in which goods and services are provided without pay or exchanged through a system of reciprocity. The public sphere is one which is dominated by economic and political exchange and tends to be associated with men. Many writers believe that even in late twentieth-century Western society, women are not fully accepted in the public arena and are allowed to participate there only on sufferance. There is an intermediate space between the public and the private which incorporates the community and neighbourhood. This can be seen as a physical extension of the family or household and is generally an area in which women are more active. The public and intermediate spaces will be developed in more depth in the next chapter, whilst the discussion here focuses on the private world of the family and household.

Table 5.2 shows a recent analysis, by the UK Statistical Service, of the division of household tasks. Even though 38 per cent of women work full time and a further 29 per cent work part time (HMSO, 1998), a disproportionately high percentage do the washing and ironing and plan the family's meals. Very few men take the main responsibility for grocery shopping and none were recorded as always or usually looking after sick family members, although they are more likely to undertake small repairs around the house (HMSO, 1996). This, it should be noted, is not quite the same as the other chores dominated by women because it is neither (a) routine (conducted regularly on a daily or weekly basis), nor (b) depended upon for survival (as is providing food and health care).

Amongst Parisian families, Jeanne Fagnani has found that, whilst 'new men' now share some of the household tasks and childcare, the organisation of this falls on the mother. This happens to the extent that women choose their jobs to be closer to home so that they can maintain contact with their children during the day (Fagnani, 1993: 178–179). Likewise, Nicky Gregson and Michelle Lowe have found in their research into the use of paid domestic workers by middle-class households in the UK, that the management of this labour is almost always the woman's responsibility (Gregson and Lowe, 1994).

Table 5.2 *Division of household tasks in the UK, 1994, by percentage*

Task	Always women	Usually women	Equally	Usually men	Always men
Washing & ironing	47	32	18	1	1
Deciding what to have for dinner	27	32	35	3	1
Looking after sick family members	22	26	45	0	0
Shopping for groceries	20	21	52	4	1
Small repairs around the house	2	3	18	49	25

Source: HMSO, 1996.

This division of tasks is more pronounced in non-Western households as Table 5.3 shows for a rural area in Colombia. Here, women are more likely to prepare meals for the family and farmworkers and do the laundry, as well as feeding hens and pigs and fetching water. Men were more likely to be engaged in gardening, milking, agriculture and cattle work.

Table 5.3 *Female/male ratios of adults (over 12) who 'always participate' in tasks, Colombia*

Tasks	San Lucas	El Distrito	La Payoa
Workers' meals	43.0	7.1	16.1
Laundry	10.1	8.1	18.1
Family meals	6.1	8.1	18.1
Feed hens	3.1	8.1	11.1
Fetch water	1.1	4.1	3.1
Feed pigs	1.1	7.1	15.1
Fetch wood	1.6	4.1	3.1
Gardening	1.6	1.3	1.2
Milking	1.25	1.7	1.1
Agriculture	1.60	1.52	1.70
Cattle work	0.65	1.28	1.13
Total population	1.1.9	1.1.5	1.1.3

Source: Townsend, 1993: 150

Katz's work in Sudan likewise shows how men (with the help of older boys) are responsible for animal husbandry and agriculture, whilst 'nubile' women, confined in purdah to the compound, are responsible for food preparation, laundry and milking. Children, teenagers and older women collect wood and water. Anoja Wickramasinghe demonstrates in Table 5.4 how women in rural Sri Lanka undertake the bulk of chores in household maintenance and child-rearing; men are more likely to work in the field and forest. Figure 5.1. displays quite clearly the number of hours women spend 'in work' which is higher than those spent by men. Particularly notable is the complete absence of leisure time for women.

In Upper Volta, West Africa, one of the few time budget surveys undertaken[2] (Figure 5.2) shows that women not only undertake more of the subsistence tasks (women's total daily work time is 587 minutes compared to men's 453), but that they also have considerably less leisure time than men (women spend 235 minutes on personal needs, compared

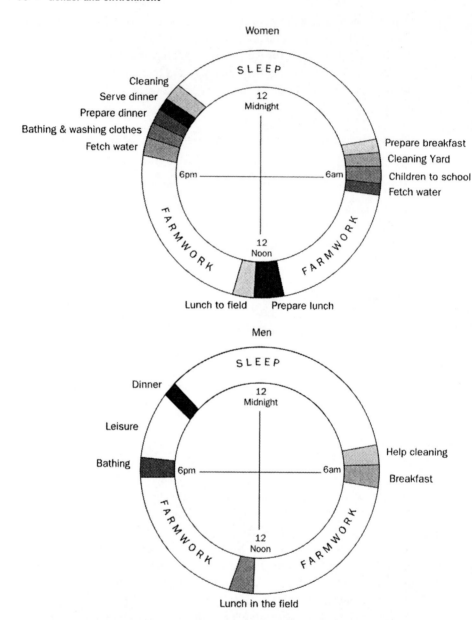

Figure 5.1 *General patterns in the time allocation of men and women, Sri Lanka*
Source: Wickramasinghe, 1997.

to 387 by men). It is clear to see how such social roles are constructed when boy and girl children's involvement in work is considered in Table 5.5 (McSweeny and Freedman, 1982). From the age of seven, girls are involved in considerably more work hours each day than boys. One of

Table 5.4 *Gender roles in household maintenance, Sri Lanka*

Task	Percentage of hours		
	Men	Women	Jointly
Food purchase	30	23	47
Decision on diet	0	73	27
Food preparation	0	93	7
Cooking	0	100	0
Preserving food	0	87	13
Serving meals	0	97	3
Cleaning kitchen	0	100	0
Washing dishes	0	100	0
Fetching water	0	70	30
Gathering fuelwood	0	53	47
Sweeping house	0	93	7
Sweeping yard	37	37	26
Washing clothes	10	70	20
Attending to children	8	61	31

Source: Wickramasinghe, 1997: 129.

the reasons given for this is the unequal opportunity given for schooling, where families are less likely to send girl children to school for a formal education.

One important point to bear in mind is the division of tasks into 'productive' and 'reproductive' activity. The first implies that the work is paid and that as a result the wage transaction is recorded in each country's system of national accounting. This work is generally (although not always, as in the case of home-working) undertaken in the 'public' sphere. 'Reproductive' tasks are unpaid, and consequently are not recorded in a country's system of national accounting. Because these tasks are universally more likely to be undertaken by women, as the tables have shown, women's work is more likely to be invisible in national accounts. More than this, the

Table 5.5 *Comparison of phasing into workloads, Kongoussi Zone, Upper Volta, by sex*

	Average hours of work daily	
Age	Girls	Boys
7	5.3	0.7
9	7.4	2.8
11	8.5	3.2
13	6.0	5.2
15	8.8	4.4

Source: McSweeny and Freedman, 1982.

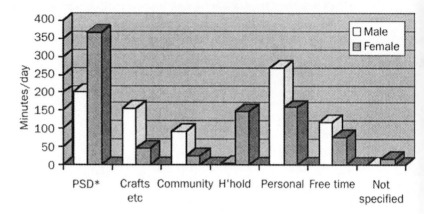

*Production, supply, distribution

Figure 5.2 *Women and men's time budgets, Kongoussi Zone, Upper Volta*
Source: Derived from McSweeny and Freedman, 1982.

United Nations System of National Accounting (which is imposed on all member countries) considers 'primary production and the consumption of their own produce by non-primary producers [to be] of little or no importance' (in Waring, 1988: 63).

Confusingly, the same activity may be considered both productive (recorded in Gross Domestic Product – GDP) and non-productive (use value): a family may grow its own food, a mother may care for her children and cook meals for her family and this work, because it is not paid, is considered 'of little or no importance', does not enter the cash economy and fails to contribute to GDP. However, in a family which buys all its own food, pays for childcare and eats in restaurants, each of these transactions is recorded and contributes to GDP. A society which relies on the latter is more likely to be considered to have a high standard of living, because the standard of living is measured by GDP, than one which relies heavily on unpaid work.

Selma James reports on historical surveys which have attempted to quantify the value of housework by estimating the monetary value of the hours spent on it. In 1970, housework in the USA was estimated at 37 per cent of Gross National Product (GNP), and in Canada, 1978, at 40 per cent. Various US companies have valued women's work at: $257/week (Chase Manhattan, 1972), $283/week (Prudential Life, 1977) and $14, 500 a year (Cornell University, 1980). In West Germany in 1984, women's contribution to the national economy was estimated to be DM6, 000m ($3, 500m) per annum (James, 1995). In Latin America, women's work has been valued at $240/month (Chile, 1983) and 7, 363m pesos in 1978 (Mexico), whilst the total contribution by housewives in Argentina would come to $21, 000m/year or 33 per cent of GDP (Ahooja-Patel, 1985 in James, 1995).[3]

Marilyn Waring, an ex-Member of Parliament from New Zealand, has written passionately and cogently about the inequity of women's uncounted work and her argument can be used to show how such an elevation of paid work over unpaid has a negative effect on the environment. Often, the activities which are valued most in Western society are those which have the greatest impact upon the environment, whilst those which have the least value leave the faintest imprint. The result of this is that, because economies seek to maximise their 'standard of living' as measured by GDP, environmentally damaging activities are encouraged at the expense of activities which have less impact.

Waring gives the example of the Exxon Valdez oil spill off the Alaskan coast in 1989 in which the damage directly contributed to the GDP of Canada and the USA by way of detergents bought, people hired to clean up the area and legal action brought (Waring, 1995). On the other hand, a pristine or relatively clean environment of itself has no monetary value and therefore does not contribute to GDP.

Mothers who feed their babies with formula or commercially bottled milk are directly contributing to their country's GDP, whilst those who breastfeed for longer periods do not. There is therefore some incentive for economies to encourage mothers to feed their children with commercially produced milk, even though evidence suggests that this has negative health and developmental effects on the child (Waring, 1988: 168–171). The famous case of Nestle which exported powdered milk to Africa, packaged with instructions in English, demonstrates how this can have serious health and environmental repercussions: powdered milk needs to be mixed with sterilised, that is boiling, water. Even had women been able to read the instructions on the packets, for many securing a regular supply of boiling water would be difficult as the discussion in the section below will show.

As Chapter 7 will illustrate, countries in the South are under increasing pressure from global institutions to conform to a standardised system of accounting. Loans and other financial help are often contingent on recipient countries agreeing to adjust their economies to the Western model. Such structural adjustments can put a considerable burden on the household and community, specifically on women whose unpaid work is central to the survival of these units.

Because the division of labour between men and women places them in quite different relationships with the environment and exposes them to different environmental impacts it can be said that environmental impacts have a gendered dimension, and this will be considered next.

The gendering of environmental impacts by household roles

This chapter has primarily been concerned with women's and men's roles within the private arena of the household. (Chapter 6 will look at roles within society, although the two spheres, public and private, are closely linked.) It has demonstrated that women are more likely to be involved in activities that, whilst essential to survival, involve no cash transaction and have, accordingly, little status. These roles include: cleaning, laundry, food preparation and cooking, grocery-shopping, childcare and care of other dependants. In the South, these tasks are augmented by: growing food for the household, fetching water and, sometimes, collecting firewood. Each of these roles has implications for the environment but, more importantly, is affected by various forms of environmental manipulation and degradation. Looking at women's roles as reproducers and consumers, I shall examine one of these activities to explore how and why gender and environment are inextricably linked.

Food: growing, harvesting, shopping, preparing, cooking

Growing and harvesting

In 1985, a United Nations, Food and Agriculture Organisation (FAO) study established that women were responsible for 80 per cent of the world's agricultural production (Dankleman and Davidson, 1988). Subsistence agriculture is almost exclusively a woman's task and without this work, her family would not eat; however, women have title to only 1 per cent of the world's land. Intensive agriculture relying on pesticides and hybrid seeds is claiming land previously used for subsistence (and frequently degrading it in the process through erosion and desertification) which displaces subsistence agriculture to less fertile areas. Subsistence farmers (the majority of whom are women) have to travel further to get to land and must work harder to try to restore it.

Traditionally, women have been the repository of much knowlc concerning seeds and soils, because of their roles in subsistence agriculture. As development agencies move into Third World countries, this knowledge is frequently ignored by development workers who directly approach men to negotiate the uptake of new agricultural practices, such as cash-cropping of high-value goods (such as mange tout peas, avocados and asparagus). Alternatively, the knowledge is 'bought' and patented by the agro-chemical industry which makes it illegal for women (or anyone else except the patent holders) to continue using the full diversity of seeds once at their disposal. This has significant impacts on their ability to farm effectively. The role of transnational companies and international agencies is examined more closely in Chapter 7.

In the West, farming is predominantly a male-dominated industry. However, a recent article suggests that organic farming is much more attractive to women. In the UK, whilst only 5 per cent of farmers using chemicals are women, almost 50 per cent of organic farmers are women (Poklewski Koziell, 1999). This, however, still represents a small overall percentage of farmers since only 0.5 per cent of UK land area is under organic management (Browning, 1999).

Water and fuelwood collection

Most rural households in the South rely on water collection and this is almost exclusively a concern of women and children (Dankleman and Davidson, 1988; see also Tables 5.3 and 5.4). The World Health Organisation estimated that, in 1980, 70 per cent of rural populations in Kenya, Tanzania and Angola had little or no access to safe water. Dankleman and Davidson report that some women spend up to four hours a day collecting water and transporting it on their heads (1988: 32). If water is not clean there is a likelihood of infection and there is, therefore, a need for sterilisation by the boiling of water. However, most rural households in the South rely on biomass for heating and cooking and this, too, is getting scarcer. As forests are cleared for infrastructural development, agriculture, dam construction and mining, firewood collectors (primarily women) are having to travel further. In the Gujarat Plains of India, it is not unusual for women to spend four to five hours a day collecting firewood, when once they would have done so once every four to five days. In addition, women are carrying increasingly heavy loads of up to 35 kilograms, 15 kilograms more than the limits many countries officially put on load-bearing for women (Dankleman and Davidson, 1988), and greater than that laid down in the International Labour Organisation Maximum Weight Convention, 1967 (Anker, 1997) (Figure 5.3.). In refugee environments, an additional burden is placed on women, who continue to attempt to fulfil the domestic roles they would normally undertake at home. In addition, refugees primarily comprised women and children (80 per cent, according to one recent newspaper report, De Groot, 1999). It has been estimated that refugee women and girls have had to travel, on average, one and a half hours per household per day on top of the journey they would normally expect to make to collect water and firewood. This generates an extra 20, 000 women hours/year (Black, 1998: 113). As wood runs out, it is increasingly substituted by dung, straw, husks and roots none of which burn as well as wood and which produce more hazardous smoke.

Cooking

Women in the South are, exclusively, the cooks and can spend an average of four hours a day cooking. Using the substitute fuels mentioned above is more hazardous, but even cooking with firewood is a health hazard as the cook inhales an amount of benzopyrene (a poisonous gas from burning fuel) which is the equivalent of inhaling the smoke from twenty packs of cigarettes a day (Dankleman and Davidson, 1988: 72). As Chapter 4 has shown, those who cook are also more vulnerable to respiratory and eye diseases.

Shopping

In the West, food is mostly purchased, as opposed to grown, by the household. Since women are still most likely to do the grocery shopping, as well as prepare and cook the food (see Table 5.2), they are most directly concerned with food-safety issues, such as irradiation, genetic modification and use of pestcides. Many environmental campaigns have recognised this and have targeted women to be mindful of what they buy. Box 5.1 lists the food, and other health/environmental campaigns, that the Women's Environmental Network (WEN) has been involved in since 1990. WEN was set up in the UK by Bernadette Vallely as a splinter organisation from Friends of the Earth which, she felt, was not sufficiently addressing issues of concern to women.

Figure 5.3 *Indian women collecting biomass*
Source: Photograph by Richard Kevern.

Box 5.1

Women's Environmental Network campaigns

Campaign	Brief description
Tampon alert	To declare Toxic Shock Syndrome a notifiable disease.
Sanitary protection disposal	To stop flushing of sanitary products as this pollutes water supplies.
Chorine phase-out	Thought to cause endemetriosis (menstrual disease).
Dioxin phase-out	Concern over health problems, especially in children.
Breast cancer	Concern that this is caused by environmental factors, e.g. increased use of lindane, a pesticide.
Packaging	Returning packaging to the manufacturer.
Reusable nappies	Encouraging mothers to use terry rather than disposable nappies.
Eco-labelling	Lobbying for accurate and comprehensive eco-labelling.
Chocolate	Wants testing for pesticides and better working conditions for producers.
Breastfeeding	Encouraging this rather than relying on commercial products.

As well as environmental danger to humans, the consumer (in this case, primarily women) may also be aware of the environmental impact of products she is buying, such as chemicals in cleaning products and sanitary wear, non-recycled paper or over-packaged goods. Surveys which investigate purchasing habits repeatedly show that women are more likely to buy environmentally sensitive goods (and to recycle the containers) than men (Buckingham-Hatfield, 1994). In a society which tends to confer the rights of citizenship on consumers, the purchase of goods is one area in which women could be seen to have some potential influence on environmental decision-making (see Lister, 1996).

In a popular book on downshifting, Ghazi and Jones claim that 'Shopping is far from a trivial business. It is . . . about making political choices. We have a responsibility to get it broadly right. It's your opportunity to change the world and yourself' (1996: 170). However, there is a danger that, by focusing too heavily on women's power as a consumer and role as the family's housekeeper, she is charged with the responsibility for a global clean-up (Figure 5.4) which obscures the true nature and causes of environmental problems, which were explained in Chapters 2 and 3. This has also been found to be so in Venezuela where

Figure 5.4 *Woman nurses the Earth*
Source: Ontario Advisory Council on Women's Issues.

the country's Ministry of Women instructs/educates women community leaders in the skills needed to improve the quality of life in settlements. Maria-Pilar Garcia-Guadilla argues that this serves to reinforce women's responsibility for environmental conservation (Garcia-Guadilla, 1993).

Where women have challenged their 'typical' gender role by entering paid work, particularly in higher-paid, professional occupations, they have often only been able to do so by employing other women to care for their children (if they are mothers) and household (Gregson and Lowe, 1994). This clearly removes these women from a more intimate involvement with certain aspects of daily reproduction, even though it is still largely the women's responsibility to manage domestic labour. It is, of course, mainly women who are employed in the domestic labour force. Food prepartion and provision is only one area in which women take prime responsibility. Try undertaking a similar exercise for childcare, or for cleaning and laundry. Given that women are the prime unpaid performers of these tasks, how are they likely to be more concerned about environmental problems, and what environmental problems might this role expose them to?

Summary

Although, as some of the tables in this chapter have shown, women the world over are largely responsible for the domestic tasks which I have used to illustrate the link between gender and the environment, it is important to bear in mind that the nature of this responsibility is affected by a woman's status and wealth. It is poor women who are particularly disadvantaged by environmental degradation as they are not usually in a position to delegate the household chores which expose them to environmental vulnerability.

The examples worked through here show how it is (poorer) women who are exposed to greater environmental risk because of their position in the family or household. Another argument which emphasises the connection between gender and environment stresses that because of these roles and because through them (poorer) women are brought more immediately into contact with nature, they have a greater understanding of it. This argument has been used by the United Nations in its attempt to bring women into environmental decision-making and this will be explored in Chapter 7. Whilst this is undoubtedly the case in, for example, subsistence agriculture, where women farmers have a substantial knowledge of seeds and soils (Dankleman and Davidson, 1988), there are critics of this claim. Cecile Jackson (1993), for example, argues that women, through their social position, may actually be prevented from taking more positive attitudes towards the environment. Women's environmental relations are, in her argument, mediated by a range of factors such as class, and the relationship a women has with and between her families (that is, the one she was born into and the one she marries into). Jackson maintains that a woman's conjugal contract can mean that she actually has less emotional investment in the land she works, since it is not legally hers.

This chapter has suggested that women are exposed to particular environmental problems, and have a particular relationship to the environment, because of the division of labour in the private sphere of the household. Although the form of the family – household changes through time and space, the role of women as prime carer of dependants and provider of cooking, cleaning and food-collection services is surprisingly persistent and universal across cultures. This suggests that there are pervasive, fundamental forces which structure such divisions of labour. Although higher income permits some women to delegate these tasks, thereby bringing them into the public sphere (for example, by paying other people to cook and clean, or by buying ready prepared meals, see Gregson and Lowe, 1994), this does not change the overall gendered structure of household labour.

The nature of this labour involves women (especially those with low income or involved in a subsistence economy) in a particular relationship with nature which exposes them more directly than men to the externalities of environmental degradation. Although there is general agreement on this argument, more controversial is the claim that some writers make for women's empathy with nature because of these roles.

Increasingly, international (and national) agencies are seeking to involve women in environmental programmes both to utilise their knowledge and to ease the environmental burden they bear. As this, and the following chapters explore, there is an inherent danger in focusing exclusively on women's roles, as this can too easily naturalise them and work against women trying to break out of this mould. A fine line needs to be drawn between respecting and utilising women's knowledge derived from the roles they have played and consigning them for ever to these roles.

Gender roles in the home are inextricably linked to the roles men and women hold in the public arena: in work, in politics and the community.

Chapter 6 will review how these roles are played out and will examine the implications of this for the environment.

Discussion questions

1 To what extent does your family fit the 'nuclear' family, which is still, against the evidence, seen as the 'norm' in Western society?

2 Assess the gendered balance of unpaid work undertaken in your current household, and in your family home – i.e. which work is done by which household members?

3 Using an example similar to food, such as childcare, cleaning or laundry, examine how women's part in these activities both exposes them to environmental problems and increases their awareness of environmental issues.

Further reading

Dankleman, I. and Davidson, J. (1988) *Women and Environment in the Third World: Alliance for the Future*, London: Earthscan. This was co-published by the International Union for the Conservation of Nature and reflects its practical concerns. As well as pointing out the hardships which many women in the Third World endure, the book also identifies a number of case studies which demonstrate women's initiatives to improve social and environmental conditions, such as the Green Belt, a tree replanting initiative in Kenya. It is written in a very easy to read and accessible style, and, although over ten years old, continues to be a classic book on women, environment and development.

Katz, C. and Monk, J. (1993) *Full Circles: Geographies of Women over their Life Course*, London: Routledge. This is an edited collection of papers, by women, on the nature of women's lives from a range of different countries. It gives an idea of the shared problems facing women as well as the differences between women from different cultures.

Waring, M. (1988) *Counting for Nothing: What Men Value and What Women Are Worth*, Wellington, NZ: Allen & Unwin. This is an extremely powerful book written by the youngest and first woman MP to be elected to the New Zealand Parliament. Having been appalled at the undervaluation of women's work world-wide, Waring made her PhD research on the UN System of National Accounting and its effect on women. She began this work as a result of her position on her Government's Public Accounts Committee, although she resigned from the government over its agreement allowing American nuclear ships to dock in NZ ports.

Notes

1   Hayford argues that it is likely that women discovered the possibility of cultivating vegetable products and that this enabled them to create the basis of a strong locality. See also Haraway's work on primates described in Chapter 2.

2   Because these surveys require men and women to be followed by recorders of their activities for all their waking hours (noting their activities, minute by minute), these surveys are very costly in time and money. Consequently there are few undertaken.

3   The first steps towards national recognition of unpaid work have been taken in Canada where the census now asks a question regarding the number of hours respondents spend looking after children, the home and caring for others without pay. The UK government has accepted a question on carers which is being tested for the 2001 census (Barber *et al.*, 1997).

6 ► Gendered roles in the community

Key words: wage differentials; feminisation of poverty; planning profession; women and politics; grass roots activities; environmental decision making; Agenda 21; citizenship

- Gender and power
- The gendering of environmental protest
- Environmental decision making

Just as women and men perform different roles in the private space of the home, so do they in public space. Activities outside the home are dominated by paid work, but also include unpaid work in the community and all kinds of political activity. The previous chapter has shown how, in capitalist society, activities are valued according to their financial rewards: housework and care of dependants within the home are unpaid and tend to be valued lightly by society. Those positions which are valued most highly generally are paid more highly: consider the status accorded doctors rather than nurses, or barristers compared to legal secretaries, or university professors compared to primary school teachers. It is universally the case that in each of the highly esteemed, better-paid professions, men dominate, whilst in those sectors where pay and esteem are lowest, women dominate. In individual professions, there are fewer women than men in higher-paid, more senior positions. In a society in which power is frequently aligned with money, this has significant consequences.

This chapter will consider how power is gendered through employment and political representation and what the environmental implications of this are. Broadly, these implications are twofold: environmental 'bads' or externalities tend to be disproportionately experienced by those on lowest incomes, whilst those on the lowest incomes are least able to challenge or influence the environmental situation. Consequently, those people least able to contribute to the environmental debate are those most likely to be negatively affected by environmental problems. This is not only affected by gender, but by class and race, age and ablebodiedness also, and the literature on environmental justice is preoccupied with disparities between North and South and, within the North, between the white population and people of colour (see, for example, Guha and Martinez-Alier, 1997 for a discussion of this).

Gender and power

In the UK, 35.3 per cent of all women of working age work full time (an average of 40.6 hours a week), 26.5 per cent work part time (an average of 18 hours a week) and 4 per cent are classified as unemployed. Almost 29 per cent of women are presented as economically inactive (HMSO, 1998). The main reason for the disparity between men and women in this category is that people (most likely women) who are financially supported by a partner cannot technically be considered unemployed. Table 6.1 compares the economic activity figures for women and men in the UK.

In the UK, women in full time paid work earn 72.7 per cent of a man's average weekly wage (£264 compared to £363). Whilst this percentage appears to have been rising since the 1970s, a recent UK analysis shows a widening of the gap in the weekly earnings of women and men in

Table 6.1 *Economic activity of men and women of working age in the UK, by percentage, 1997*

	Men	Women
Economically active:		
In employment:		
Full time	58.8	35.3
Part time	4.8	26.5
Self-employed	12.8	4.7
Others in employment (e.g. training schemes)	1.1	1.2
Unemployed:	7.0	4.1
Economically inactive:	15.5	28.8

Source: HMSO, 1998.

Table 6.2 *Women's earnings as a percentage of men's by sector in the UK, 1997*

Sector	Women's wage as a percentage of men's
Manufacturing	72
Wholesale/retail	74
Health and social work	76
Hotels and restaurants	78
Public administration/defence	79
Education	88
Transport/storage	90

Source: Social Trends, 1998.

1997/98, indicating the fragility of equal opportunities in earnings (Denny, 1998). In the 'professional occupation' group, women in full-time employment earn a gross weekly income of £421 which is 90 per cent of the average wage for both sexes (Office for National Statistics, 1998).[1] Table 6.2 breaks this down by sector, in which it can be seen that in no sector do women earn more than 90 per cent of the average male wage. This wage differential can also be seen in the USA, where the median weekly earnings for women in full-time paid work was $399 in 1995, compared to $522 for men (US Bureau of the Census, 1995). The world average female: male ratio for all non-agricultural work is 77.8 for the hourly rate, 76.7 for the daily/weekly rate and 71.6 for the monthly rate (Anker, 1997). Table 6.3 breaks this down into a sample of countries' wages recorded by the International Labour Office, in which only one country, Swaziland, is found to have a ratio showing women to earn more than men. (The hourly ratios show the best comparison as some women work shorter weeks and months than men.)

Women are more likely to work in public services than men and any pay squeeze in this sector will be felt disproportionately by women. The *Guardian* newspaper reports Alistair Hatchett of Income Data Services accusing the UK government of gender discrimination in pay, particularly through public-sector wage restraint for nurses and teachers (Denny, 1998). Repeatedly, doctors' salaries, though also paid out of the Treasury, are not subject to the same restraints.

This income profile combined with lone parenthood (in the UK 1 per cent of men are lone parents with dependent children, whereas 5 per cent of women are; HMSO, 1998) and a poorer old age (women are less likely to have work-related pensions because they have either not been in paid employment at all or have been in part-time or discontinuous employment), constitutes what is known as 'the feminisation of poverty' whereby women are more likely than men to be poor and the greatest proportion of people in poverty are women. Research in Canada has,

Table 6.3 *Female/male earnings ratios in the world, around 1990*

Region/country/area	Reference period	Female/male wage ratio	
		All non-agricultural	*Manufacturing*
OECD countries			
Australia	Hourly	88.2	82.5
Belgium	Hourly	75.1	74.5
Denmark	Hourly	82.6	84.6
Finland	Hourly		77.3
France	Hourly	80.8	78.9
Germany (Fed. Rep.)	Hourly	73.2	72.7
Greece	Hourly		78.4
Iceland	Hourly	87.0	
Ireland	Hourly		69.2
Luxembourg	Hourly	67.8	62.2
Netherlands	Hourly	77.5	75.0
New Zealand	Hourly	80.6	74.9
Norway	Hourly		86.4
Portugal	Hourly	69.1	69.0
Sweden	Hourly		88.9
Switzerland	Hourly	67.9	68.0
United Kingdom	Hourly	70.5	68.4
Cyprus	Daily/weekly	59.0	58.0
Turkey	Daily/weekly	84.5	81.0
Developing countries/areas			
Sri Lanka	Hourly	91.2	
Egypt	Daily/weekly	80.7	68.0
Hong Kong	Daily/weekly	69.5	69.0
Sri Lanka	Daily/weekly	89.8	88.0
Costa Rica	Monthly	66.0	74.0
Japan	Monthly	49.6	41.0
Kenya	Monthly	78.3	73.0
Korea (Rep. of)	Monthly	53.5	50.0
Malaysia	Monthly		50.1
Paraguay	Monthly	76.0	66.0
Singapore	Monthly	71.1	55.0
Swaziland	Monthly	106.6	88.0

Source: ILO Yearbook of Labour Statistics, in Anker, 1997.

likewise, shown that women form the majority of Canada's poor, are the majority in social housing and on the waiting lists for social housing (Federation of Canadian Municipalities, 1997). Chapter 2 has already demonstrated how science and science education is dominated by men, especially in the more senior positions, and has considered the implications of this regarding setting the environmental agenda.That women are concentrated in the lower grades of jobs can be illustrated in two other areas which have a significant impact on the environment: planning, and politics.

Planning

In the UK, planning is a professional field of activity which requires five years of training to qualify for membership of the Royal Town Planning Institute (RTPI); RTPI accreditation is necessary for any entrant to the profession. Clara Greed's research into how strongly the planning profession is gendered has established that only 18 per cent of full members of the RTPI are women and that this percentage of women practitioners is high compared to the proportion of women in other professions concerned with the built environment such as architects, surveyors and civil engineers. Of all professionally qualified staff in local authority planning offices in England and Wales, 23 per cent are women.[2] However, only 2 per cent of Chief Planning Officers and 5 per cent of Deputy Chief Planning Officers are women. Therefore, in a field which has a considerable impact on the environment, women are very poorly represented, particularly in policy making positions (Greed, 1994).

That this situation is unlikely to be significantly challenged in the near future is confirmed by other European research which echoes this pattern of expertise in the teaching of the built environment professions. In a sample of Schools of Architecture in Europe, it was found that on the faculty, women held 5 per cent of professorships in Germany, whilst in Greece the average was 10 per cent. Such profiles are unlikely to challenge the way in which architecture is taught, nor to encourage many more women to enter this profession. In the UK, 25 per cent of students are women, whilst in Germany, this figure is 39 per cent (European Commission, Equal Opportunities Unit, 1994).

As will be explored later in this chapter, this is likely to have implications for the nature of decisions made, including the neglect of aspects specifically relevant to women. Planning in the UK is closely governed by elected members at both the local and national level and here, too, representation is heavily male-dominated.

Politics

In the political sphere, women have low representation and very little power. Representation follows a pyramidal structure similar to that of planning, as suggested above, and science as discussed in Chapter 2. Twenty seven per cent of local councillors in England and Wales are women (LGMB, 1998) as are 18 per cent of UK Members of Parliament.[3]

In the European Union, 25 per cent of MEPs are women (the *European Companion*, 1998), as well as 25 per cent of the commissioners. (Forty five per cent of European Commission staff are women but only 13.5 per cent of those holding 'A' grade jobs, European Commission, 1994.) Table 6.3 shows the degree of female representation in parliaments in 1987 and 1990 in the OECD countries.[4] The Scandinavian countries had the highest proportion of female representation. Interestingly, not all countries have shown an increase in female representation between 1987 and 1990, indeed some have recorded a decline.

In Central and Eastern Europe, the situation has been reversed in the past ten years. Under communist regimes, women had greater political prominence, facilitated by widespread nursery provision and an expectation that women worked outside the home. In Hungary, for example, many social services have been cut since 1989 and more women have been 'pushed into the background in every form of political participation' (Koncz, 1994: 5). In 1975, one-third of the members of the Hungarian Parliament were women; by 1994 this had dropped to 6.7 per cent, with no women ministers, two state secretaries and three under-secretaries of state (Koncz, 1994: 9).

Of course, simply increasing the numerical representation of women in politics, planning or any other organisation will not necessarily make a difference to the substantive issues dealt with, or the processes involved. Programmes such as Opportunity 2000 in the UK are involved in equal opportunities of employment, rather than changes in decision-making itself. Generally, women entering typically male-dominated fields are required to fit into a pre-existing structure and often have to achieve

Table 6.4 *Female representation in parliaments (lower houses), by percentage*

	1985 (UN data)	1990 (Janova & Sineau)
Austria	11.5	10.9
Belgium	7.5	8.5
Denmark	29.1	30.7
Finland	31.5	31.5
France	6.4	5.7
Germany (Federal Republic)	15.4	20.5
Greece	4.3	4.3
Iceland	20.6	20.6
Ireland	8.4	7.8
Italy	12.9	12.8
Luxembourg	14.1	13.3
Netherlands	20.0	25.3
Norway	34.4	35.8
Portugal	7.6	7.6
Spain	6.4	13.4
Sweden	28.5	38.1
Switzerland	14.0	14.0
UK	6.3	6.3
Canada	9.6	
USA	5.3	
Japan	1.4	
Australia	6.1	
New Zealand	14.4	

Source: Lane, McKay and Newton, 1997.

higher goals than men to be considered for promotion (see, for example Greenfield's newspaper article in Chapter 2). Interestingly, Robinson quotes research which suggests that women do better, with regard to pay, under incentive schemes based on results rather than subjective evaluation (Robinson, 1998). However, it is possible to find examples of organisational structures which have been changed by women. In Australia, Judith Matthews found that in Adelaide the planning office, which was primarily staffed by women, had adopted a more cooperative way of working which had permitted the planning process there to be more responsive to public (including women's) participation. Arguably, the Labour government in the UK, which came to power in 1997 with a record number of women MPs, has been able to challenge the overt masculinity of the Houses of Parliament in which there was a shooting range but no creche, a barber but no hairdresser and which conventionally debated bills from 2 p.m. until the early hours of the morning because members could then attend their Boards of Directors' meetings or law practices in the morning. Under the leadership first of Ann Taylor and subsequently Margaret Beckett, some of these practices are being changed to bring the House of Commons more in line with late twentieth-century workplace practices.

In Canada, a move has been made to extend the participation of women in municipal government. In 1993, the Quebec City municipal council assigned half the seats on its executive committee to women and in 1996, a regulation was adopted making it compulsory that seats be reserved for women on neighbourhood councils. (Nevertheless, in the higher echelons of provincial politics, only 9 per cent of Quebec's mayors were women and 20 per cent of municipal councillors elsewhere in the province were women.) Similar procedures safeguarding a percentage of seats for women can be found in India, Brazil and Tanzania (Federation of Canadian Municipalities, 1997). There is a considerable debate around the creation of women-only positions. Favouring the development is the argument that because women are discriminated against, if not directly, then by the prevailing structural factors which discourage, say, flexible working and maternity leave, unless women-only seats are created, women will never have an opportunity to achieve positions of political power in equal proportions to men. Those opposed to quotas argue that if women prove themselves to be up to the task required, they will be selected and to create vacancies based on a person's sex/gender may deny other people who may be better suited for the position. Some women also resent the notion that their position may be thought to have been achieved on the basis of their sex/gender rather than their abilities. (It is interesting to note, however, that the quota system has been used in the USA to ensure more equal opportunities for African Americans, with some success.)

Non-governmental organisations

In non-governmental organisations (NGOs) it might be expected that women would have a larger role. They are, after all, much more likely to be involved in grassroots community action. Nevertheless, as Joni Seager points out, beyond the grassroots, women are much less prominent. She estimates that in the USA, 60–80 per cent of grassroots activity is undertaken by women and yet there are virtually no women in senior positions in environmental NGOs. In 1989, the woman president of WWF was the first to head a major environmental membership organisation (excluding specifically women's organisations).[5] Seager writes that no women had ever held a top post in the leading environmental organisations until then, and that only around 30 per cent of upper management personnel in these organisations are women. She also goes on to report how poorly ethnic minorities are represented in US environmental organisations with 0.9 per cent in the Audubon Society, 0.4 per cent in the Sierra Club, 3.6 per cent in the Natural Resources Defence Council and 12 per cent in Friends of the Earth, US (Seager, 1993). Ethnic minorities constitute over 20 per cent of the US population. As NGOs become increasingly successful in lobbying government, such under-representation further marginalises groups with little say in the environmental debate.

The disparity between grassroots activity and environmental group leadership has also been noted in Latin America by Garcia-Guadilla (1993). In Venezuela, men are more likely to lead environmental groups, with the exception of, until recently, neighbourhood groups involved in quality of life issues. Yet, even since the 1960s, women have participated more in organisations that demand public or collective urban services such as parks, water, schools and pavements (Ray, 1969, in Garcia-Guadilla, 1993). However, since 1989, the number of neighbourhood groups led by men has increased in response to a Neighbourhood Participation Regulation which has transferred political power to neighbourhood associations. Garcia-Guadilla suggests that this can be explained by the route into local and municipal politics it provides men and the political parties to which they are more likely than women to belong (Garcia-Guadilla, 1993: 84).

This section has illustrated how women have less value than men in Western and Western-influenced societies, how they are more likely to be poor and less likely to be in positions to influence decision-making. The environmental justice debate shows how particular groups of people suffer environmental problems more acutely because they lack a political voice against those who make decisions which have significant impacts on the environment, and because of their inability to buy their way out of an area which is polluted. For example, in 1987, the United Church of Christ Commission for Racial Justice published *Toxic Waste and Race in the United States* which reported that 58 per cent of African-Americans, and 53 per cent of Hispanics, lived in communities where the dumping of hazardous wastes is uncontrolled (Merchant, 1996: 60-61). In 1994, by signing an Executive Order on environmental justice, President Clinton acknowledged that communties of colour and low income receive a disproportionate share of polluting industries and waste sites. This provides guidance to Federal and state agencies to examine whether such communities are being deliberately targeted by polluting industries. It also has the power to consider whether clean-up operations vary between different socio-economic groups (Gibbs, 1998: 8).

The next section will consider how women, often with low incomes, have tried to challenge industry and government to mitigate environmental problems. Using the examples of Love Canal in USA, the Chipco movement in India and the Women's Peace Camp at Greenham Common, UK, I will explore how women are motivated to protest because of their vulnerability, their situation and their concerns and how these concerns are treated by the media and decision makers.

The gendering of environmental protest

The beginning of the environmental movement is usually given as the late nineteenth century, ignited by such luminaries as John Muir in the United States and John Ruskin in the UK. Whilst the key figures in this movement tend to be male, there is a long tradition of women working in the conservation movement. Carolyn Merchant in her essay on women in the progressive conservation crusade argues that women transformed the conservation crusade in early twentieth-century USA from 'an elite male enterprise into a widely based movement' (Merchant, 1996: 109).

One reason given for this transformation is that the traditional role of women as carers of home and family (see Chapter 5) is extended to protect the resources which contribute to this. Early US examples of such protection include lobbying for a School of Forestry to teach protective management; successfully fighting to set aside native redwood trees in California (Sepervirens Club); The General Federation of Women's Clubs, founded in 1890 which was active in forest conservation; and the Women's National Rivers and Harbours Congress (founded 1908) which fought for water conservation, the preservation of Niagara Falls and for clean shores and streams. Some of the women who founded these organisations saw their activity as an extension of their housekeeping;

Mrs Overton Ellis, speaking to the 1909 Conservation Congress called conservation:

> the surest weapon with which women might win success. Centuries of turning last night's roast into hash, remaking last year's dress, and controlling the home's resources had given women a heightened sense of the power of the conservation idea in creating true womenhood. Conservation in its material and ethical sense is the basic principle in the life of a women.
>
> (Quoted in Merchant, 1996: 128)

Whilst it is important not to let this argument become deterministic in the sense that conservation is seen as 'natural' to women, it can be seen how women, forced into a housekeeping role, might extend this housekeeping beyond the home to protect key resources such as water and forests. However, in tune with Seager, Merchant notes that as the conservation movement became more professionalised, the women's involvement in this was marginalised. In the Venezuelan context, Garcia-Guadilla has argued that by continuing to identify the environment and environmental problems with the domestic sphere, government is less likely to address these problems. Whilst women appear to be successful in raising broader environmental issues which fall outside the household, such as the protection of a national park, to the attention of government, in Venezuela, they have a marked lack of success in politicising those environmental issues which are seen as being in the realm of the household (Garcia-Guadilla, 1993: 81-82).

In the modern environmental movement which can be dated from the early 1960s, women are still the main grassroots activists, galvanised by issues such as reproductive health and the health of children and other family members as affected by such things as radioactive and hazardous waste, pesticides and herbicides and nuclear weapons and power. Chapter 4 pointed out the negative impacts of these pollutants on the body. However, it is difficult for these women to get their voices heard as the example of the Love Canal Homeowners Association protest in Niagara demonstrates.

Lois Gibbs, a housewife living in, as she put it, a 'middle-class, blue collar' neighbourhood, became concerned about pollution when she noticed her son's health deteriorating after he enrolled at a school which had been built over a toxic waste dump. Frustrated by the school board's refusal to allow her son to be transferred, Gibbs canvassed local support to close down the school. As the extent of the neighbourhood's health problems became clear, the Love Canal Homeowners' Association (LCHA) was formed and run by local women (Gibbs was the Chair). The LCHA collected health data through interviews with residents and used this to fight for the resettlement of three hundred households affected by the dump. One of the problems for residents (and, arguably a contributor to the strength of the protest) was that all their capital was tied up in their houses. As house prices fell dramatically as news of the toxicity was broadcast, residents could not afford to move out of the area to buy elsewhere, as they would not be able to recoup their investment. Gibbs's story is an indictment of how government at various levels refused to accept the local data as valid, famously referring to it as 'housewives' data'. In the short term, the LCHA fight was successful in that eventually all affected households were offered compensation which allowed them to move, but not before a disproportionately high number of cancer-related and perinatal deaths had occured.

In the longer term, however, the fight was only partially successful in that resettlement was granted in the 'interests of mental stability' of the residents (that is, not acknowledging the environmental health problems) and that part of the area was later 'revitalised' for poor families to be moved into with Federal loans, since banks would not lend on the properties. Gibbs's own conclusion is that the state serves vested interests (which do not include 'housewives') and is only responsive to pressure at critical moments preceding elections (Gibbs, 1998). But one outcome she did identify was the transformative effect of the two-year-long protest on the lives of women who had participated in the protest, resulting in greater freedom and independence, although often at the expense of a marriage. She also suggests that this sustained protest challenged traditional roles in that, as women became more active in social action, their husbands were left to take on a greater share of domestic tasks. Similar transformations have also been noted in Venezuela (Garcia-Guadilla, 1993), Sri Lanka (Wickramasinghe, 1995/1996) and Bangladesh (Hashemi and Morshed, 1997) as Chapter 7 will show in the context of women's increasing economic and political involvement.

At the same time (late 1970s and 1980s) on the other side of the world, in India, women had been fighting to save part of the Himalayan forest, around Nahi-Kala village, which was being destroyed by mining activities. Not only was this causing deforestation, but water sources were also drying up. Women became involved in peaceful protest (such as laying down in front of trucks) because 'without our freedom, and forests, and foods, we are nothing, we are impoverished' (Shiva, 1993:

249). Chapter 5 has already shown how women have prime responsibility for the collection of fuel wood and water in India. In a recent publication, the success of the Chipco movement has been attributed to the charisma of the (male) leaders: Prasad Bhatt and Sunderlal Bahuguna. Ramachandra Guha designates Bhatt and his environmental organisation, DGSM as the 'pioneer of Chipco' (Guha and Martinez-Alier, 1997: 154). However, in Vandana Shiva's interview with women Chipco participants, she sets out to explode '[t]he myth that movements are created and sustained by charismatic leaders from outside . . . as non violent struggle . . . in which ordinary women like Itwari Devi and Chaumundeyi have provided local leadership through extraordinary strength' (1993: 246).

Whilst women were not the only protestors at Nali-Kala or Love Canal, they are the ones who initiated and sustained the protest at the grassroots. Their initial motivation was to protect resources necessary to nourish the household (the Chipco movement) and to protect their families from hazardous and toxic chemicals (Love Canal Homeowner's Association). Both protests have become part of the iconography of the environmental and women's movement, but it is edifying to examine how these myths develop, and how the processes are described by different commentators.

The final example I will present of an environmental movement concerns the establishment of a women's peace camp at Greenham Common, a US nuclear weapons base in the UK. Women are generally considered to be more opposed to nuclear power/weapons than men. For example, Chapter 2 summarised PRISM's survey on women in science education which suggests that the nuclear defence bias in physics research is offputting to female students, whilst Merchant quotes research that claims that 'a much larger percentage of women [58 per cent] than men [41 per cent] considered themselves definitely or somewhat anti-nuclear (Merchant, 1996: 151).

In 1981, the Greenham Peace Camp was established at Greenham Common, Berkshire, UK, a US cruise missile base, by a group of women from South Wales. These self-styled 'women for life on earth' marched from Cardiff, Wales, to Greenham to protest against the UK government's stance of military aggression and stockpiling of nuclear weapons. Whilst the initial camp was supported by men and women, in 1982 the organising group decided to make the camp women-only in order to create a strong message of feminine cooperation and non-hierarchy to counterpose the maleness of the highly combative and hierarchical military base. They also thought that a women-only camp would prevent any violence between protestors and police which might be provoked by male presence on the camp. This fear was partially realised when some of the departing men left, though the violence was inwardly directed towards the women's shelters (Cresswell, 1994: 37).

Since then the camp has attracted much publicity which mostly castigated the women for being derelict in their womanly duties such as to be clean and fragrant, cook well and be houseproud. It is an irony that action taken by the women to protect the family, community and home (Blackwood, in Cresswell, 1994: 56) has been interpreted with such hostility (Pettman, 1996: 112). Nevertheless, such publicity kept the issue of nuclear warheads in the public eye and, it could be claimed, has had some impact on increasing support for weapon decommissioning in the 1990s.

There has been a long tradition of women's involvement in the peace movement. At the end of the First World War, one thousand women attended the Hague Peace Conference to appeal for peace and argued that women could not be protected under conditions of modern warfare. The group urged a humane peace process to supplant what they believed was a vengeful peace agreement at Versailles which divided Europe up into imperfect ethnically determined states and decimated German territory, an act which others later believed contributed to the rise of Hitler and the Third Reich. At the time, a women's international conference held in Zurich presciently claimed that this 'would create discords and animosities which can only lead to future wars' (Women's International League for Peace and Freedom, in Pettman, 1996: 108).

There are many more examples of women's peace movements in places as diverse as Israel, Chechnya, Argentina, Chile, Ireland and Yugoslavia (see for example, Cynthia Cockburn's (1998) book on the rapprochement forged between women across ethnic divides in Northen Ireland, Israel/Palestine and Yugoslavia). In the latter, the women's lobby is campaigning against a nationalist, chauvinist government and has set up an SOS hotline for women and children victims of violence which reports rapidly increasing violence within families, including the use of guns (ibid.: 129).

Chapter 2 has already shown how Western society is grounded in a (Baconian) philosophy which values an abstraction from nature, allowing society (or at least a controlling sector of society which is highly masculinist) to control both nature and less powerful members of that society. Links are frequently made between the environment and peace movements in both of which women are prominent at the grassroots. This can be linked to the feminist arguments in Chapters 2 and 3 which suggest that women are more inclined to be peaceful or non-violent either because of their social roles (socialist eco-feminist arguments) or their nature (essentialist).

In summary, women are active at the grassroots of environmental protest. Their commitment to this protest is due in part to their social roles of housewife and mother, although some commentators argue that the 'emotional' nature of women empowers them to lead and take positive decisions in this area. (See Helen Caldicott's work quoted by Merchant, 1996: 154 and the discussion on cultural eco-feminism in Chapter 3.) Environmental protest can also be seen as an extension of community activity in which women have been more traditionally involved, partly because of their family responsibilities and their confinement to 'private' space. There is a wide literature on the link between the 'everyday life' of housework and caring of family members and the 'intermediate space' of communities and neighbourhoods (between the private space of the home and the public space of paid work and politics) which argues that women are tied to the latter by the former. (See, for example, the rich Scandinavian literature on 'everyday life' represented by Horelli, 1995; Horelli and Vepsa, 1994; Vepsa and Horelli, 1995.) This is particularly the case for poorer women (see the discussion on the feminisation of poverty, above) who are often tied to environmentally undesirable places and have to devise strategies for coping with problems as they are unable to utilise the option to move, as the Love Canal case study illustrates.

Nevertheless, in undertaking environmental protest, it is not uncommon for women to be underestimated or vilified or for their protest to go unrecognised. As the Chipco movement illustrated, in the formalisation of environmental protest, it is usually men who assume positions of responsibility and power. Thus the environmental movement itself, as well as reactions to it, is profoundly gendered. The final section in this chapter will consider how the environmental decision-making process is gendered and will look in particular at a relatively new instrument influencing environmental decision-making in localities world-wide: Agenda 21.

Environmental decision-making

This chapter has shown how both politics and planning are gendered professions or activities. As a result, at the location where environmental decisions are made, women's input is minimal. Women have been most active in private and intermediate spaces and yet, as environmental protest becomes more professionalised through non-governmental organisations, it assumes traditional corporate structures in which men and masculine practices dominate. Other factors, such as corporate donations to NGOs, also mitigate against true grassroots representation in making decisions in these organisations.

NGOs are becoming increasingly important in setting the international environmental aganda, as the next chapter will show, and the Women's Environmental Network has shown the potential for women's involvement in the preparatory committees which preceed United Nations agreements. In an attempt to raise the participation of all disadvantaged groups in environmental decision-making, the United Nations agreed Agenda 21 at the Rio de Janeiro Conference on Environment and Development (UNCED) in 1992, under pressure from the NGO lobby, a significant achievement. Signatories committed themselves to depositing a national plan for sustainable development by 1994 and local areas were required to produce local strategies by the end of 1996 (since revised). One of the guiding principles of Agenda 21 is that people normally excluded from the decision-making process (women, indigenous people and young people, for example) need to be integrally involved in decision-making within a framework which stresses the importance of public participation. The reason for this inclusive form of participation is that these underrepresented groups are seen as having had little impact on the production of environments, although they are sometimes disproportionately affected by them.

Agenda 21

Chapter 24 of the United Nations Commission on Environment and Development's Agenda 21 (United Nations, 1992) requires signatories to Agenda 21 to raise the capacity of women to participate in environmental decision-making and to ensure that structures of decision-making facilitate this. This entails a wide programme to promote literacy, health care, childcare, equal opportunities, the ending of discrimination in both the public and the private spheres, and the portrayal of women in a positive way. The document recognises that women and children are particularly vulnerable to environmental damage and their perspectives need to be incorporated in environmental decision-making by the active and equal participation of women at all levels. Given this unequivocal commitment, it is instructive to consider how governments have responded in their submissions to the United Nations outlining how they plan to achieve sustainable development.

In the UK *Strategy for Sustainable Development* there is one mention of women – in the section on non-governmental organisations. Even in sections which discuss population, household formation and income any gendered perspective is missing. Stressing its commitment to the nuclear family, 'The Government believes that *couples* should make their own decisions about how many children to have' (HMSO 1994: 8; my emphasis); this presupposes that all parents are part of a 'couple' and marginalises woman's right to determine her own fertility. In his introduction to the document, the then Secretary of State for the Environment, John Gummer, opened his statement with 'Man has grown used to living as conqueror' (HMSO, 1994), a statement which well illustrates the legacy of 'Enlightenment' thinking, described in Chapter 2. In the 1996 *Common Inheritance* document, participation is not mentioned at all for women. There is, as a consequence, nothing in the commitment tables about women (or participation) in this UK document. This tendency has been continued under the Labour government, whose consultation paper, 'Sustainable Development: Opportunities for Change', makes no reference to marginalised groups. Whilst the document expands on the Conservative government approach to sustainable development to combine strategies to achieve sustainable development with those to redress social exclusion, there is no reference to which groups are most vulnerable to exclusion. Neither does the document address this issue in sections on participation or consumption (Department of the Environment, Transport and the Regions, 1998).

There is an observable trend that government (local as well as central) is transforming citizens into consumers, but even in the more protracted discussions on consumers, there is no acknowledgement that consumption activity may be gendered. This is interesting as elsewhere women are repeatedly framed in terms of consumption. Monk and Garcia-Ramon (1996), for example, show clearly how citizenship in Europe is linked to economic independence and the effects this has on women's access to resources.

Andrews cites Ruth Lister who believes that 'active citizens' are those 'able to stand alone, independent before the market, their freedom guaranteed by economic rather than social rights', suggesting a prioritisation of the consumer (Lister in Andrews, 1991: 13). The exercise of rights and responsibilities also depends on information. In some cases a lack of information may prevent people from playing a full part in civic life, in others it may be that the information and knowledge that people have to offer is not recognised by decision makers, as in the case of Love Canal. Elsewhere, Lister (1996) argues that citizenship itself has a gender dimension, suggesting that, by its emphasis on participation in public life, women are often excluded from full citizenship. Reinforcing this is the ambiguity of rights in the private sphere and the 'right' to citizenship exercised through participation in paid employment. The same argument can be extended to children and to others not in paid employment and confined to the private sphere, such as many of the elderly and people with disabilities.

Judith Matthews's analysis of the Australian response to Agenda 21, the *National Strategy for Ecologically Sustainable Development* (ESD) identifies a concern with gender issues with regard to the particular concerns of women and any inequitable effects on women as a result of government programmes. State governments were encouraged to take action to ensure women's access to and awareness of information and decision-making processes, and to recognise women's interests in the development of these processes. State governments were also expected to seek input from women's representative groups, ensuring women's participation in cooperative approaches to resource management issues, and ensuring participation of women on ESD – related decision-making bodies and advisory groups. In the 1995 Annual Report of the Australian Commonwealth Government to the United Nations Commission on Sustainable Development, the section relating to women notes that 'Australian women see environmental issues as a major concern'. However, the report notes that this importance, and the high profile that women have in environmental community groups and in the debate on environmental issues 'is not yet adequately reflected in many of the decision making processes which have significant impact on the environment' (Commonwealth of Australia 1995: 54). The report goes on to outline various undertakings to improve participation by women, and notes the changing emphasis on 'gender' issues in work by development agencies, although no subsequent government statements have 'women' mentioned in their title.

Although the response of the two Western governments has been markedly different in responding to Agenda 21, it is, as yet, too early to estimate the outcome of this. Reviews of Local Agenda 21 (that aspect of planning for sustainable development at the local level) in the UK suggest that the procedures for involving traditionally marginalised groups in the decision-making process are not dissimilar to procedures used in previous planning exercises (Buckingham-Hatfield and Matthews, 1999). As such they are unlikely to generate qualitatively different kinds of decision-making. This would be a useful area to explore in a range of locations, and would be an interesting dissertation research topic.

In a recent survey of all UK local authorities, only 7 per cent had actively encouraged or monitored women's involvement in Local Agenda 21, despite women being a UN priority group (Buckingham-Hatfield and Riglar, 1999). Indeed, whilst research in West London has demonstrated that women's concern (and particularly women with children) for environmental issues is consistently higher than is men's, this has not been pursued in the Local Agenda 21 programme which has been primarily driven by the planning profession (mostly male officers) and by male-dominated groups such as the cycling lobby and by state agencies, all of which were represented by men. (Buckingham- Hatfield 1994, 1999).

Summary

Environment is experienced differently by men and women as a consequence of the different daily 'worlds' in which they operate. This difference in experience constitutes a significant element of what has been referred to as *place identity* – that element of a person's sense of self which derives from their experience of, and attitudes towards, the physical environment. Val Brown argues that women's uses of the environment are sufficiently different from those of men as to constitute a distinctive 'habitat', in the ecological sense (Brown, 1992). Such a position echoes that proposed by Mary Mellor (1996) in arguing that women are more embedded in the environment, as a result of their greater involvement with the work of nurturing and caring. This results in an awareness of the impacts of environmental damage which needs to be acknowledged and developed by both women and men in the process of working towards sustainable development.

This chapter has considered how men and women participate in different spatial and qualitative areas of activity: private, public and intermediate; paid and unpaid work. Because of the values of Western society which elevate the public and the paid over the private and unpaid, this has an impact on those people (more likely to be women, as well as children, the elderly, people of colour and disabled people) involved in less valued work.

The specifically environmental impact concerns:

1 Those with undervalued and unrecognised contributions to society who are more likely, because of their lack of financial wealth, to suffer environmental problems which they cannot easily escape.
2 These people are least able to do anything about these environmental problems because of their lack of a recognised public role, which effectively confers on them second-class citizenship.

It is important to recognise the continuum between the most private space of the body to the wider public spaces of paid work and political representation as environmental inequalities which arise in one space have impacts along the rest of that continuum. This chapter has concluded with the local and national implementation of an international programme to see if the rhetoric of environmental justice is likely to translate into material differences to the lives of people I have been describing in this book. The next, and last substantive chapter will consider the impact of international organisations and of global economic and social processes on women and men in different communities and countries, with a specific emphasis on the environment related initiatives of the United Nations and the World Bank.

Discussion questions

1 Taking any occupation or profession that you are familiar with, analyse the gendered distribution of jobs and pay within it.
2 Examine any environmental protest movement to see who is active in this at the local level, and who makes decisions concerning the direction of the protest. Do you find this to be gendered?
3 Consider your local area's Local Agenda 21 programme and see who is participating in this, and in what capacity. Are there areas which may be of concern to women generally, or particular groups of women, which are missing?

Further reading

Breton, M. J. (1998) *Women Pioneers for the Environment*, Boston, MA: Northeastern University Press. This is an inspiring collection of cameos of women environmentalists, both contemporary and historical, mainly, though not exclusively, from North America.

Buckingham-Hatfield, S. and Matthews, J. (1999) 'Including Women: Addressing Gender', in Buckingham-Hatfield, S. and Percy, S. *Constructing Local Environmental Agendas*, London: Routledge. Much of the material on LA21 in the UK and Australia is covered in more depth in this chapter.

Cockburn, C. (1998) *The Space Between Us: Negotiating Gender and National Identities in Conflict*, London: Zed Books. Here Cynthia Cockburn writes about her participant observation of three women's groups in Northern Ireland, Palestine/Israel and Yugoslavia which are trying to bridge the divide between conflicting groups. As well as being a highly readable account of women's attempts at forging peace, it contains her photographs of the process.

Gibbs, L. (1998) *Love Canal: The Story Continues*, Gabriola Island, BC, Canada: New Society Press. This is an extremely readable first-hand account of a neighbourhood's fight against toxic waste. As well as telling of the difficulty and frustration of dealing with local and state government, Lois Gibbs writes about the way in which community activity empowers women and potentially changes their lives.

Greed, C. (1994) *Women and Planning: Creating Gendered Realities*, London: Routledge. This is the fullest description of the work of women in the planning profession, written in an easily accessible way by a Reader in Planning at the University of the West of England. Clara Greed has also written a number of planning textbooks in which the role of women features prominantly, and books on women's experience in other built environment professions.

Merchant, C. (1996) *Earthcare*, London: Routledge. This book has been recommended several times already, but a number of chapters give a good account of the women's environmental movement in the US, Australia and Sweden.

Notes

1 The Labour Force Survey, from which this data are drawn, does not give the average gross weekly salary for men in this sector. These statistics are for winter 1996/97.

2 Qualified planners may also work in the private sector where pay tends to be higher. The disparity between women members of the RTPI and women

working in local authority planning offices suggests that qualified women
planners are more likely to work in the public sector.

3 In 1998, 23.4 per cent of Labour MPs are women, 8.5 per cent of
Conservative MPs, 6.5 per cent of Liberal Democrat MPs and 33 per cent of
Scottish National Party MPs. Other parties have no women representatives.

4 The OECD stands for the Organisation for Economic Cooperation and
Development which represents the world's most developed countries.

5 Interestingly, one of the USA's major environmental organisations, the
Audubon Society, was founded by a woman, Harriet Hemenway, a wealthy
Boston socialite who was outraged at the decimation of heron rookeries
because of the fashion for feathers in women's hats. She drew on her social
register colleagues for support to form the Massachusetts Audubon Society,
which became a model for other state Audubon societies. These were later
federated into the National Audubon Society (Breton, 1998).

7 The international dimension

Key words: international division of labour; structural adjustment; micro-credit; gender and development; United Nations

* The international political economy
* Gender and investment
* International interventions

> [T]he eradication of poverty is one of the fundamental goals of the
> international community and the entire United Nations system . . . and it
> is essential for sustainable development. Poverty eradication is thus an
> overriding theme of sustainable development for the coming years . . .
> Eradication of poverty depends on the full integration of people living in
> poverty into economic, social and political life. The empowerment of
> women is a critical factor for the eradication of poverty.
>
> (United Nations, 1997, in Osborn and Bigg, 1998)

Globally, women constitute around half the population (although as
Chapter 4 has shown, the number of women is less than could be
expected through the natural pattern of birth and survival in infancy).
They account for one-third of the *official* workforce, yet do two-thirds of
the productive work. You will remember from Chapters 5 and 6 that
much of the work that women do is not counted in national accounts or
in the United Nations System of National Accounting. Women earn
one-tenth of the world's income and own less than 1 per cent of its
property (Pettman, 1996).

Such impoverishment is the result of processes at different scales. Local
processes are responsible for geographical differences in the relationship
between men and women – for example, particular cultures regulate
marriage conventions, child-bearing and rearing practices and access to
education in particular ways. The differences in household structures
between parts of the Caribbean (see Pulsipher, page 66 in this volume)
and the Sudan (Katz, page 66) have already been noted and these

structures are very different from the way in which households are organised in the West. This chapter is more concerned with fundamental global processes which structure gender relations and which either, in turn, or in parallel, affect the environment. I will first consider the impact of economic processes on gender and the environment through the international division of labour and the global economy which are both driven by the capitalist model of society which originated in the West. Secondly, I will look at political responses to these impacts, particularly through the activities of the United Nations, but also of international financial organisations, such as the World Bank, and of non-governmental organisations.

The international political economy

Just as there is a local division of labour at the household, community, regional and national scale, so is there at the international scale. Although much of women's work is not counted because it takes place outside the formal economy (see Waring, 1988, and Chapter 6) and involves no payment, a gender division of paid labour is also noticeable at the global scale. Increasing numbers of women are entering the paid labour force in the developing world, although as a percentage of women, it is lower than in the West (Stichter, 1988). Capital exploits regional differences by tending to locate enterprises where labour and other cost savings can be made; globally this involves the location of plants where a balance of skills and cheap labour can be found. The geographical transfer of low-skill manual-assembly jobs is closely associated with the increase of women in the manufacturing labour force (Stichter, 1988). The wage discrepancy between the North and South is considerable, and is particularly marked in the case of women and children. Table 6.3 has already shown how persistent is the gap between male and female earnings world-wide where in all but two of the countries listed, the gap was larger in manufacturing. Women's wages in this sector fell to around half of men's in several of the developing countries: Malaysia (50.1), Republic of Korea (50.0), although the largest gap was in Japan (41.0). One-third of female–male pay differentials are estimated to be due to the sex bias in occupational sectors (Anker, 1997). This kind of *horizontal segregation* sees women and men concentrated in different sectors. For example, Anker suggests that men, world-wide, are more likely to be prominent in primary-sector jobs; that is, where there is better pay, security, working conditions and opportunities for advancement. These jobs tend to be with firms which have labour-market power which insulates them from competition, to a large degree. On the other hand, women are more likely to be concentrated in the secondary sector, where pay, chances for promotion, working conditions and job security are poor. Firms in this sector tend to face fierce competition (Anker, 1997). The latter characterises firms in export processing zones, which overwhelmingly employ women and children (Pettman, 1996). In addition, Stichter argues that women in the developing world are more likely to be employed in the service industry (70 per cent of all female labour is hired in this sector of which 39 per cent of its labour force is women, 1988: 331). Women are also disadvantaged in paid work because of *vertical segregation* through which women are more likely to hold the lower-paid jobs in any one sector. This is compounded by the unlikelihood of women being offered training as industrial processes become more mechanised, at which point the number of men employed rises (Stichter, 1988: 331).

Ansara has claimed that 'Capitalism and globalised development policies are worked on the bodies of women' (1992, in Pettman, 1996: 197). Whilst this could be said to be true of many workers who, for example, suffer from poor health and safety regulations (see the discussion of the Union Carbide chemical plant which exploded in Bhopal in Chapter 4), it is particularly true of women not only because of the double or triple day that they may work (see Chapter 5), but also because of the increasingly lucrative trade of sex tourism, through which the bodies of women and children are frequently advertised as a 'natural resource' (Pettman, 1996).

The trade in women is not restricted to prostitution but also extends to the sale of brides to Western men (Humbeck, 1996 and Pettman, 1996) and to the export of domestic staff to expatriots abroad. In this practice, women leave their country to work for households to which their visas are tied. Because they may not leave this household without losing their work permit, and because they will not generally be able to afford their fare home, this effectively converts a (poorly) paid position into servitude. Cases which have been brought to the attention of the host country frequently cite the sexual exploitation of women employed in domestic positions. States may encourage these forms of activity since the wages women send home to their families provide valuable foreign exchange. Between 1 and 1.7 million Asian women are thought to be in domestic service overseas, whilst the Philippines receives an estimated US$3 billion a year in remittances from overseas workers (Pettman, 1996: 189–191).

As well as being sexual victims to the international economy, war also exploits women. Records of the Cambodian conflict recorded the number of prostitutes rising from 6, 000 to 20, 000 within twelve months of Americans arriving (De Groot, 1999). This is in addition to the increasing exposure of women to military conflict as refugees (see Chapter 5), victims of military rape and as direct casualties (see Chapter 4). This escalation is taking place at a time when the military itself is increasingly distancing itself from combat, at least in international wars. Such tactics (for example the remote sensing of targets) incidentally also result in greater environmental damage, as witness the recent war in Kosovo.

As the military increasingly takes on an international peacekeeping role, the United Nations is recognising the value of diluting the masculinity of the army with women peacekeepers, for a number of reasons. De Groot (1999) reports recent American experience which suggests that males are more likely to control their sexual urges in mixed units, whilst the largest percentage of war victims (women) are more likely to respond to another woman. De Groot (1999: 5) quotes the UN Division of the Advancement of Women calling for a 'critical mass' of women in all future operations, arguing that: 'Without women's participation there can be no real progress in the resolution of on-going conflicts'. This is in the light of 'notable UN successes where there was a greater than normal female presence' (that is, in Guatamala, Namibia and South Africa).

Women are also particularly affected by structural adjustment policies which are imposed by global institutions (for example, the World Bank) to 'streamline' economies to which they lend (that is, to make them more cost effective). The public sector is generally the first to be affected by these structural adjustment policies, as services are cut and shifted to the household and the community. Women are also most likely to make those 'invisible re-adjustment[s]' when social security expenditure and food subsidies are abolished (Pettman, 1996: 168). They may do this by becoming more involved in the informal economy by selling home-made goods or food, selling 'personal' services, including prostitution (Pettman, 1996: 168) or by taking in lodgers (Garcia-Ramon, 1996).

As Garcia-Ramon's work illustrates, it is not just women in the South whose labour is sought by global enterprises. In Europe, the differential wage rates between the Mediterranean countries and North West Europe account for increasing numbers of women in Greece, Portugal and Spain entering the low-wage, formal economy, often in seasonal, irregularly paid work, or in home-based work. Sabate-Martinez, for example, looks at how industries have sought to minimise their labour costs by deliberately locating in large villages which have a large population of women without other employment alternatives. These women may never before have had paid employment. In addition, the clothing manufacturing industry, which has exploited pay differentials between northern and southern Europe by locating in Spain, Portugal and Greece, relies heavily on homeworking which, as an employment sector, is notorious for its 'intensive use of clandestine workers, who are always women' (Sabate-Martinez, 1996: 273).

Whether in response to structural-adjustment policies or to economies adjusting to global processes, women in particular have to compensate for shrinking welfare entitlements and the shift of responsbilities from the state to the community. In extreme circumstances, strategies women adopt to attempt to shield their families from poverty involve them forgoing food, as Chapter 4 has already shown.

Gender and investment

Traditionally, investments made in environmental projects tend to favour men. Jodi Jacobson, a researcher at the Worldwatch Institute in Washington, DC, writes that women do not always benefit from economic growth. Even where standards of living (measured by standard indicators such as life expectancy, per capita income and primary school enrolment) show a rise, inequality between the sexes obdurately remains. She quotes the United Nations Development Programme (UNDP) Human Development Index as showing women's level of access to the resources needed to attain a decent standard of living as 94 per cent of men's level in Finland (the highest percentage world-wide) and 70 per cent of men's level in Kenya and South Korea (the world's greatest discrepancy) (Jacobson, 1993).

Development programmes tend to assume that what is good for the man is good for the family, but this is frequently not so as the incomes of men are more likely to be treated as personal income and spent on consumer products such as tobacco and alcohol. A 1992 World Bank report has shown how children's nutritional levels have fallen even when their father's income rises. This income has been found to be spent on watches, radios and bicycles. Children were found to have a better diet when the mothers earned an independent income (Jacobson, 1993).

Indeed, Sylvia Chant has found from her research that lone mothers are often in a better financial position than women living with husbands who spent their income on gambling, tobacco and alcohol (Chant, 1997).

Global interventions into national economies can have many different negative impacts. Wickramasinghe, writing about her native Sri Lanka, argues that large-scale, externally funded commercial agricultural and forestry ventures not only drain money from the state's economy, but also exclude women and are insensitive to the environment (Wickramasinghe, 1995/1996; 1999). Whilst

> women in agrarian societies are the backbone in the labour force in the informal rural economy. Yet, the benefits of mainstream development have not reached them, because development assets are delivered to men who are the formal heads in the patriarchal households.
>
> (Wickramasinghe, 1995/1996: 145)

However, this failure to recognise the value of women's contribution cripples efforts to achieve broad development goals and may be an ineffective strategy. The Grameen Bank, the first micro-credit organisation, which originated in Bangladesh in the 1970s, has found women to be the most reliable and conscientious borrowers, as Box 7.1 shows.

As well as securing the loans, such a lending strategy has been found to be more effective in reducing poverty and enhancing 'households' capacity to sustain their gains over time' (Hashimi and Morshed, 1997: 223). Other improvements which have been noted include the increased nutritional status of children and higher calorific intake. Through becoming involved in this lending project, women have become more empowered through negotiating with bank officials and as they become physically more mobile between villages, by attending the necessary meetings. An unanticipated outcome of this form of credit which has been observed has been the increased use of contraceptives, linked to the length of time a woman has participated in the Grameen Bank, which has been noted (Hashimi and Morshed, 1997: 224). It should be noted that this is a voluntary use of contraceptives, rather than a result of targeted family planning programmes initiated by Western development agencies, referred to in Chapter 4.

In Europe, Sabate-Martinez has also noted a fall in fertility rates as more women are involved in paid work in Spain. She believes that this is due to changing gender relations which follow entry into the paid labour

---

Box 7.1

The Grameen Bank

The Grameen Bank was formed to lend money at commercial bank rates to the poorest of households who would otherwise have to rely on the usurious rates of loan sharks. It began as an experimental project in rural Chittagong, Bangladesh, initiated by a professor from Chittagong University. It now receives its funding from the Bangladesh Central Bank, local commercial banks and international donors. Ninety-four per cent of the Grameen Bank's borrowers are women and the bank has a repayment rate of 98 per cent. Loans are made for individual projects but are issued only to groups of four to five women who take joint responsibility for the loan repayment, which encourages group support and unity. These groups manage the loan and get involved in the community economy through regional federations. The initial loan of $US75 to $US100 is made to the two neediest members of the group who must repay the loan before any other members of the group receive a loan. The loan may be made for any purpose which is approved by the group and the centre chief (each centre is responsible for a number of groups). The loans are repaid in fifty equal installments, and when this is repaid, the borrower may also apply for a larger loan. Members are required to save money which is then converted into Grameen Bank shares. Before receiving a loan, group members are expected to undertake training which helps them to manage the loan and their new business.

Source: Summarised from Hashemi and Morshed, 1997.

---

force (Sabate-Martinez, 1996). Wickramasinghe, too, has observed the positive outcomes which emerge when investments are made direct to women, such as a more secure food supply, better clothing, improved housing conditions and more money spent on children's education.

Whilst the Grameen Bank was a locally generated enterprise, in Sri Lanka it was a Norwegian-funded development group which helped to set up these women's economic projects which have now been incorporated into a national Integrated Rural Development Programme. Eighty-three per cent of projects have been successful and Wickramasinghe's evaluation estimates that 30 to 40 per cent of families have moved out of poverty through investing loans in 'home-based enterprises'. Whilst Wickramasinghe stresses the importance of home-based enterprises as they fit into women's established sphere of activity, she also notes an increase in women's self-confidence as an outcome of this programme (Wickramasinghe, 1995/1996). Whilst it is important to bear in mind that investments into household and community based enterprises may reinforce women's roles and prevent them from challenging these, both programmes have observed that by lending to women, their status in the household rises as they claim entitlement to participate in decision-making. The women's experience has also been found to increase community recognition of women and they are increasingly filling political positions in the villages (Wickramasinghe, 1995/1996: 146).

Such programmes which raise education levels of women, and thereby, children, are critical in developing social and environmental equality. However, they are far outweighed by the conventional lending of the World Bank, and even non-governmental organisations, who, Rahman claims, do not accept that the poorest 5 to 10 per cent of households are reliable borrowers (Rahman, 1997: 274). Nevertheless, micro-credit has shown itself to be effective and is now being explored in the West as a possible strategy for poor, women-headed households.

There is, however, increasing pressure on lending agencies to adopt a 'gender and development' approach to environment and development issues. This requires a broad consideration of the relationship between men and women. When such considerations are made, it is possible to compare the effects of directing investment to women and men. In addition to the examples of micro-credit given earlier, the World Bank has noted that 'women who are trained to manage and maintain community water systems often perform better than men because they are less likely to migrate, more accustomed to voluntary work and better entrusted to administer funds honestly' (World Bank, 1992). However, as is illustrated by a case in Mali, West Africa, women's input into World Bank-funded projects is just as likely to be the provision of food for association meetings (Joekes *et al.*, 1996).

The 'gender and development' approach is gradually replacing the earlier 'women in development' approach which typified the way in which women's concerns were addressed in environment and development projects in the 1960s and 1970s. This tended to 'bolt' women's concerns on to existing policies, rather than to acknowledge the complex social interactions between women and men. As an example of good gender and development practice, Joekes *et al.* give an example of a water and sanitation project in Casmance, Senegal, which successfully trained women in the assembly and maintenance of hand pumps (Joekes *et al.*, 1996). Such technical training, along with the micro-credit initiatives presented earlier, not only take on board women's concerns and roles, but begin to challenge these roles and empower women. In turn this can have significant effects on gender relations in the local area.

Of all international institutions, the United Nations is probably the most advanced in addressing gender differences in environmental issues, as the next section will demonstrate.

International interventions

The Fourth, United Nations-sponsored, World Conference on Women in Beijing in 1995 prioritised the following issues for women: access to education; improving literacy rates; improved health services (especially for pregnant women); improved family planning and information on contraception and sexually transmitted diseases; combating poverty and the marginalisation of women and food security (Pietila and Vickers, 1996: ix).

The United Nations argues that there is a close relationship between the status of women and the state of the environment. It particularly signals education as a critical factor which will have a long-term impact on improving women's status, and on reducing societies' impact on the environment (particularly with regard to population growth). The two 'micro' programmes summarised in the previous section have illustrated how education and the empowerment of women may have this effect. This belief underpinned the Earth Summit held in Rio de Janeiro in 1992 (the United Nations Conference on Environment and Development), particularly the agreement of Agenda 21[1] which incorporated a chapter on women. Chapter 24 of the Agenda 21 document required the countries which signed up to this commitment to bring women into environmental decision-making for two reasons. Firstly, they are disproportionately affected by negative environmental impacts because of their social and domestic roles, and greater likelihood of poverty. Secondly, because of these roles, they have a closer relationship with, and knowledge of, the environment, whether as farmers, food and meal providers or primary health care providers. Box 7.2 contains the objectives and activities which signatories to Agenda 21 are expected to pursue to achieve the greater equal opportunity and participation of women.

The last Secretary General of the UN, Boutros Boutros-Ghali, believed the United Nations, as an institution, had to lead by example and he supported the setting of a goal to bring the gender balance in United Nations policy level positions as close to fifty-fifty as possible by the end

Box 7.2

Chapter 24, Agenda 21

Global action for women towards sustainable and equitable action – objectives

The following objectives are proposed for national Governments:

1 to implement the Nairobi Forward-Looking Strategies for the Advancement of Women, particularly with regard to women's participation in national eco-system management and control of environmental degradation;
2 to increase the proportion of women decision makers, planners, technical advisers, managers and extension workers in environment and development fields;
3 to consider developing and issuing by the year 2000 a strategy of changes necessary to eliminate constitutional, legal, administrative, cultural, behavioural, social and economic obstacles to women's full participation in sustainable development and in public life;
4 to establish by the year 1995 mechanisms at the national, regional and international levels to assess the implementation and impact of development and environment policies and programmes on women and to ensure their contributions and benefits;
5 to assess, review, revise and implement, where appropriate, curricula and other educational material, with a view to promoting the dissemination to both men and women of gender-relevant knowledge and valuation of women's roles through formal and non-formal education, as well as through training institutions, in collaboration with non-governmental organisations;
6 to formulate and implement clear governmental policies and national guidelines, strategies and plans for the achievement of equality in all aspects of society, including the promotion of women's literacy, education, training, nutrition and health and their participation in key decision-making positions and in management of the environment, particularly as it pertains to their access to resources, by facilitating better access to all forms of credit, particularly in the informal sector, taking measures towards ensuring women's access to property rights as well as agricultural inputs and implements;
7 to implement, as a matter of urgency, in accordance with country-specific conditions, measures to ensure that women and men have the same right to decide freely and responsibly the number and spacing of their children and have access to information, education and means, as appropriate, to enable them to exercise this right in keeping with their freedom, dignity and personally held values;
8 to consider adopting, strengthening and enforcing, legislation prohibiting violence against women and to take all necessary administrative, social and educational measures to eliminate violence against women in all its forms.

Source: from Earth Summit, 1992.

of 1997. However, as he recognised, simply bringing more women into positions of power is not, of itself, enough: male roles also need to be changed (Pietila and Vickers, 1996: 151). As Cecile Jackson points out in relation to gender struggles and environmental conservation, women cannot be considered to be an homogeneous group and their relationship to the environment, and to each other, is mediated by their age, class, wealth, social roles and positions in the family (Jackson, 1993: 402). She also cautions that Western women tend to see third world women as victims which is not always helpful or welcome (ibid.: 388). It is not enough, then, for gender balances to be constructed numerically.

As yet, it is too early to estimate how successfully women are participating in Agenda 21, although, as Chapter 6 has attempted to show, there appear to be few initiatives which have fulfilled the United Nations' brief. In a 1997 report on progress on Agenda 21, there are very few mentions of gender (Osborn and Bigg, 1997). The UN, through its dicentenial Conference on Women is committed to improving conditions for women and it has set up a development fund for women, to provide direct support to development projects for women. UNIFEM was established in 1976 and promotes the inclusion of women in decision-making in all development projects.

Summary

Globally, then, the gendered division of labour concentrates women in particular sectors, occupations and grades which have become an integral factor in decisions affecting industrial location. According to Hanson and Pratt, the gendered division of labour is part of the industrial restructuring process, and not just incidental to it (1995: 10). Commercial and institutional interventions world-wide have changed the activities of women, bringing them into the (low) paid labour force through industrialisation and constraining or challenging the subsistence practices of agricultural women. These are frequently two sides of the same coin. Gender relations have been affected by a migratory labour force (responding to globalisation) which creates a high number of effectively women-headed households, and which dispatches young women to the city and to live overseas to earn foreign exchange through prostitution and domestic service. Gender roles are also being challenged by more positive interventions such as micro-credit, which serve to make women more independent and offer their families higher standards of living.

Gender and environment at the global level can be seen as linked victims of global capital, echoing the relationship between women and nature seen in Chapter 3. Here, you will remember, the argument of social eco-feminists was put forward to explain how women and nature share man's domination. In this reading, women and nature share a common fate, as argued by some of the eco-feminist writers summarised in Chapter 3. In some environmental movements (exemplified in Chapter 6), this perceived relationship has been utilised by women to challenge global forces, whether it be Greenham Common protesters up against the entire Military Industrial Complex or the Love Canal community fighting the Occidental Chemical Corporation and various levels of the US government. The globalisation of industry both utilises gender and environment in its search for cheap production and market share. Both environment and gender become commodities whether explicitly, for sale in tourism (in sex tourism, mostly, but not exclusively, females; and in 'nature' – sun, sea, sand and forest), or as the means of reducing production costs via cheaper labour or by externalising the environmental costs of production.

Discussion questions

1 With reference to something that you own (e.g. trainers, or a piece of clothing produced in the South), attempt to trace it back to its origins to see who may have been involved in its manufacture and how these workers fit into the global division of labour. (Christian Aid and Oxfam have both done work in this area.)

2 Looking at adverts for travel to 'exotic' destinations in the South, consider how women and nature are treated as commodities.

3 Read the reports of war zones throughout the world and note the relative proportion of casualties amongst civilians and participants. What is the likely gender impact of this?

Further reading

Pettman, Jan Jindy (1996) *Worlding Women: A Feminist International Politics*, London: Routledge. This is one of the few books written about women and international politics which gives a good (if disturbing) review of how women have a particular experience of war, nationalism and the global economy. Whilst not directly concerned with the environment, these areas are important in thinking through the relationship of gender and the environment.

United Nations (1992) *Agenda 21*, Geneva: United Nations. Most academic libraries will have a copy of this document and it is very useful way of reading about the United Nations commitment to gender and environment.

Wood, G. D. and Sharif, I.A. (1997) *Who Needs Credit? Poverty and Finance in Bangladesh*, London: Zed Books. If you would like to read more about the Grameen Bank and other micro-credit initiatives, this is an excellent place to start.

Note

1 Agenda 21 was formally agreed at the United Nations Conference on Environment and Development and constitutes a framework within which individual states should work to achieve an environmentally and socially sustainable environment. Each state which agreed to the document was required to publish its strategy for achieving this by the end of 1994 and local governments were asked to produce local plans by the end of 1996; this has since been revised to 2000. Key elements to Agenda 21 are social equality and participation by parties not normally involved in decision-making (such as women, children and young people and indigenous people). Refer back to Chapter 6 for more details.

8 ▶ Conclusions

- Explanations of gender–environment relations
- Environmental problems and gender
- Differential capacity to affect the environment
- Strategies for changing gender–environment

This book has introduced the reader to a number of interrelated issues which govern the relationships between society and the environment. Before embarking on this, the introductory chapter explained how society itself is gendered, in that relations between men and women are unequal because of a historically sustained practice which has privileged men's position over women's. The form which a society takes is clearly going to have an impact on that society's relationship with the environment, and this is the main theme which runs through the book. I have identified four components of gender–environment relations which I have used as organising principles of this book. These are:

1 explanations of how gender–environment relations have developed;
2 the gender-differentiated effects of environmental problems;
3 the differential capacity of women and men to do anything about these problems;
4 a review of possible strategies for changing this gender bias.

I will summarise each of these components in turn and consider how they are interrelated.

Explanations of gender–environment relations

Chapter 2 spent some time looking at the way in which Western science has developed and how this was a product of prevailing social relations at any particular time. Enlightenment thought, which was in debt to Greek philosophy, was informed by a belief that men were superior to women. The justification for this was that 'man' was more distant from nature, less in its thrall, and so able to observe it more rationally and unimpeded by emotions. 'Man's' abstraction from nature was considered a virtue and enabled him to establish a relationship with nature in which nature was subservient to men's needs. Conversely, 'woman' was considered to be inferior to 'man' because she was closer to nature both physically (through bearing and giving birth to children) and emotionally. This and characteristics such as intuition and compassion were seen to make woman unfit for public life and she was therefore excluded from the worlds of commerce and politics. Women have been controlled by men both in this public sphere and in the private sphere of the household through what some writers (particularly feminist writers) argue is a patriarchal system of dominance. This dominance is still in place across the world today, as witness the remarkably consistent patterns of women's low pay relative to men's, the evident division of labour in the kind of paid and unpaid jobs men and women do, and women's lack of representation in any decision-making bodies. Such a division of labour pervades almost all sectors and positions, but has particular resonance in those professions which have the greatest impact on the environment. Leaving industry aside, which has a striking imbalance of women in decision-making positions, Chapter 2 explored how the science professions were dominated by (white) men and how this was likely to have an impact on what was studied, the way in which it was studied and on the culture of the institutions in which this was studied. In turn, this was likely to have an impact on future practitioners of the subjects as women were not encouraged to join a profession which they felt excluded them. Chapter 6 developed this imbalance by examining the planning profession and politics, and finding both to be over-represented by men, and imbued with masculine values.

Eco-feminist writers have understood gender relationships in two ways. Cultural, or essentialist, eco-feminists argue that whilst women may indeed be closer to nature than men, for the reasons above, it should not follow that women are inferior. In fact cultural eco-feminists hold that the special affinity with nature which women develop because of their reproductive capabilities qualifies women to be valued more in society, rather than less. Where gender–environment relationships are concerned, cultural eco-feminists believe that woman's special understanding of, or empathy with, nature equips her to make decisions concerning the fate of the environment; they therefore demand that the hierarchy of men–women–nature be reversed, if we are to stop the degradation of nature.

Social eco-feminists feel a little uncomfortable with this essentialist position and argue that, through patriarchy, women have been kept in a subservient role which is similar to the dominance of 'man' over nature. Social eco-feminists argue that this shared experience of domination enables women to better speak for nature, since they have less of a vested interest in maintaining the status quo. That is, they have less to gain from continued dominance of nature. Chapter 3 explored the nuances of these positions and some of the critiques of them from within and outside eco-feminism.

We can then go on to look at environmental problems as, in part, a product of this unequal society in which one sex enjoys greater power than (and over) the other.

Environmental problems and gender

Many environmental problems which exist today are the by-products of an industrial system which fails to take account of the negative effects of industrial processes. Industry has so far managed to pass these negative effects off on to the community at large, rather than to internalise the environmental costs. However, the community does not experience these effects evenly. The effects of pollution and resource depletion tend to be felt more acutely by those people and communities in poverty.

Gender relations affect the way in which environmental problems are experienced in three ways. Firstly, women's vulnerability to these problems is intensified because of their biological constitution. Secondly, because women undertake particular roles in society, this exposes them to environmental problems more acutely. Thirdly, because society is unequal and women are paid less and tend to be employed in the more vulnerable sectors, their poverty exposes them to environmental problems which wealthier groups can buy themselves out of.

Chapters 4 to 7 have shown how these inequalities are felt. For example, in Chapter 4, it was shown how women's bodies are particularly susceptible to chemicals used as pesticides, but it also showed how women suffered chemical pollution which affected the respiratory system more acutely because their social roles involved exposure to hazardous cooking smoke. Chapter 5 developed some of these ideas which show how women's enforced domestic role leaves them vulnerable to environmental degradation. This can be shown clearly through the example of the women who travel four to five hours a day to collect firewood for cooking. Where forest resources are being exploited, deforestation results in the women having either to make longer journeys to find cooking fuel or to rely on alternative cooking fuels, such as dung and husks, the smoke of which is more hazardous to women as they cook.

Differential capacity to affect the environment

The United Nations has called for the greater participation of women in environmental decision-making, in recognition that they are more likely to suffer environmental problems than are men. But the UN also recognises that women have particular skills in dealing with the environment through their social roles as mothers and housewives. To date, women's primary involvement in taking action to protect the environment (and, so, the health of their families) has been confined to grassroots activity. Chapter 6 has given examples of how women are active in opposing deforestation (the Chipko movement in India), in fighting toxic waste (Love Canal, in the USA) and in protesting against nuclear weapons (Greenham Common, UK). However, the success of these movements is highly contingent on professional environment organisations taking up their case and on the acceptance of their case by the regulating authorities. As Chapter 6 has shown, this support is far from forthcoming and it is notable that not only government, but environmental organisations also, are staffed primarily by salaried senior managers who are predominantly male.

Strategies for changing gender–environment relations

Whilst this book reviews some cases where the inclusion of women in existing environmental programmes is thought to be sufficient to address the gender bias in society–environment relations (see the women in development practices summarised in Chapter 7), most of the arguments considered propose that more fundamental changes need to take place which challenge the domination of women. Commentators are divided as to whether women's existing roles need to be worked with, to ensure that they are not exposed to environmental hazards as they are fulfilling their roles, or whether the roles themselves need to be challenged if women are to reduce their exposure to environmental problems. What is also clear is that addressing environmental degradation and addressing inequality (and not just of women, but of children, old people, ethnic minorities and communities in the developing world) have to be addressed in tandem. Whilst this book is not written to provide solutions to environmental problems, it has identified ways in which writers and activists suggest these inequalities can be redressed, from localised strategies such as the Grameen Bank, through to the global strategies put forward by the United Nations. It has also reviewed the theoretical eco-feminist perspectives which advocate particular ways of resolving environmental problems through valuing women more highly. Many of the eco-feminist philosophers are also quite unlike more conventional academics in that they advocate not only a different way of thinking, but also action on how we might reach those goals. (This is also common amongst other ecologist writers such as deep ecologists and eco-socialists, see Pepper, 1996.)

This raises an important point about the use of knowledge and an appropriate point on which to end this book is to invite you to think about the line between theory and practice and the responsibility of the academic (from the undergraduate student to the professor) to contribute towards creating a fairer and more sustainable world. This was one of the underlying principles of radical geography which emerged in the 1970s, and from which feminist geography has developed. If, like me, you believe there should be no such line, I include in an Appendix a list of organisations you may like to contact.

Appendix: further information

There are many, many ways in which we can begin to apply the theories of inequalities to live situations. Outside higher education institutions there are a myriad of organisations which would be grateful for any active involvement you are prepared to give. These include Amnesty International, Friends of the Earth, the Women's Environmental Network, environmental, peace and human rights campaigns, as well as your local area Local Agenda 21 (contact your local Council for details). A good place to start to look for these groups is in your own college or university which may have a Student Community Action group, as well as specific groups like a campus-based Amnesty International or Planet 21. Some institutions encourage students to undertake community work as part of their assessed programme (CCLC in the UK and Campus Compact in the USA will give you a list of those which do). Listed below are a number of organisations which may be useful in regard to pursuing active engagement or further insight into the issues covered in this book.

Association of University Teachers Women's Section: www.aut.org.uk/groups/eops/women.htm. (Campaigns for equal opportunities for women staff in higher education.)

CCLC (Council for Community Learning and Citizenship): CSV Education for Citizenship, 237 Pentonville Road, London N1 9NJ; tel. 020 7278 6601. (This is a national organisation which coordinates university programmes in which students get involved in community-based work for which they are assessed.)

Campus Compact: www.compact.org. (An equivalent organisation in the USA.)

The CVCP's Women in Higher Education Register:
15 Princes Gardens, Exhibition Road, London SW7 2AZ; email
aregister@ic.ac.uk. (Collects, analyses and disseminates information on
women in higher education.)

The Fawcett Society:
www.gn.apc.org/fawcett/home.html. (A long standing organisation which
campaigns for equal rights for women.)

International Federation of University Women:
www.ifuw.org/index.htm. (Works internationally to improve status of
women and girls to promote lifelong education and to enable graduate
women to effect change.)

Student Community Action (SCA): Oxford House, Derbyshire Street,
London E2 6HG; tel. 020 7739 0913; www.scadu.org.uk.

UNIFEM: 304 East 45th Street, 6th Floor, New York 10017, USA; tel.
00 1 212 906 6400; fax 00 1 212 906 6705. (An office of the United
Nations with direct relevance for women's issues.)

The Women's Environmental Network:
87 Worship Street London EC2A 2BE; tel. 020 7247 3327; fax 020 7247
4740; email wenuk@gn.apc.org. (A campaigning environmental group
with particular reference to issues concerning women.)

The Women in Geography Study Group:
RGS-IBG, 1 Kensington Gore, London SW7 2AR; email rhed@rgs.org.
(This is an affiliate group of the Royal Geographical Society with the
Institute of British Geography.)

References

Ahooja-Patel, K. (ed.) (1985) 'Office for Women Workers' Questions', Geneva:
   International Labour Organisation, 10–11, in James, S. (ed.) (1995) *The
   Global Kitchen: The Case For Governments Measuring and Valuing
   Unwaged Work*, London: Crossroads Books.
Andrews, G. (ed.) (1991) *Citizenship*, London: Lawrence & Wishart.
Anker, R. (1997) 'Theories of occupational segregation by sex: an overview',
   International Labour Review 136, 3.
Bacon, F. (1620) 'Novum organum', in Abrams, M.H., Talbot Donaldson, E.,
   Smith, H., Adams, R.M., Monk S.H., Lipking L., Ford, G. and Daiches, D.
   (eds) *The Norton Anthology of English Literature*, vol 1, 4th edn.
Bacon, F. (1625) 'Of marriage and single life', in Abrams, M.H. *et al.* (eds).
Bacon, F. (1627) 'The New Atlantis', in Abrams, M.H. *et al.* (eds) op cit.
Bandarage, A. (1997) *Women, Population and Global Crisis*, London: Zed
   Books.
Barber, S., Carroll, V., Mawle, A. and Nugent, C. (1997) *Gender 21, Women and
   Sustainable Development*, London: UNED UK.
Biehl, J. (1991) *Rethinking Eco-feminist Politics*, Boston, MA: South End
   Books.
Black, R. (1998) *Refugees, Environment and Development*, Harlow: Longman.
Bradley, H. (1989) 'Gender segregation and the sex typing of jobs', in
   McDowell, L. and Sharp, J. P. (eds) (1997) *Space, Gender, Knowledge:
   Feminist Readings*, London: Arnold.
Breton, M. J. (1998) *Women Pioneers for the Environment* Boston, MA:
   Northeastern University Press.
Brown, V. (1992) *Engendering the Debate*, report prepared for the Ecologically
   Sustainable Development Working Groups by the Office of the Status of
   Women and Department of the Prime Minister and Cabinet.
Browning, H. (1999) 'Fields of vision', *Country Living* February.
Buckingham-Hatfield, S. (1994) 'Addressing environmental issues in the 1990s:
   a gendered perspective', *West London Environmental Papers* 2.
Buckingham-Hatfield, S. and Matthews, J. (1999) 'Including women: addressing
   gender', in Buckingham-Hatfield, S. and Percy, S. (eds) *Constructing Local
   Environmental Agendas*, London: Routledge.

Buckingham-Hatfield, S. and Riglar, N. (1999) 'Participation in LA21', *International Sustainable Development Conference*, Leeds, March 1999.

Bullard, R. (1999) 'Dismantling Environmental Racism in the USA', *Local Environment* 4, 1.

Butler, J. (1990) *Gender Trouble: Feminsm and the Subversion of Identity* London: Routledge.

Cadbury, D. (1997) *The Feminisation of Nature: Our Future at Risk*, Harmondsworth: Hamish Hamilton.

Campbell, Lady C. (1997) *A Life Worth Living*, London: Little Brown.

Capra, F. (1976) *The Tao of Physics – an Exploration of the Parallels between Modern Physics and Eastern Mysticism*, London: Fontana.

Chant, S. (1997) *Women-Headed Households: Diversity and Dynamics in the Developing World*, London: Macmillan.

Cockburn, C. (1998) *The Space Between Us: Negotiating Gender and National Identities in Conflict*, London: Zed Books.

Cohn, C. (1987) 'Nuclear language and how we learned to pat the bomb' in Keller, E.F. and Longino, H. E. (eds) (1996) *Feminism and Science*, Oxford: Oxford University Press.

Collard, A. and Contrucci, J. (1988) *Rape of the Wild: Man's Violence Against Animals and the Earth*, London: The Women's Press.

Commission of the European Communities (1994) *Women at the European Commission*, Brussels, Luxembourg: CEC.

Commonwealth of Australia (1992) *National Strategy for Ecologically Sustainable Development*, Canberra: Australian Government Publishing Service.

Commonwealth of Australia (1995) *Australia's National Report to the United Nations Commission on Sustainable Development on the Implementation of Agenda 21, 1995*, Canberra: Australian Government Publishing Service.

Cream, J. (1995) 'Re-solving riddles: the sexed body', in Valentine, G. and Bell, D. (eds) *Mapping Desire*, London: Routledge.

Cresswell, T. (1994) 'Putting women in their place: the carnival at Greenham Common', *Antipode* 26, 1: 35–58.

Daly, M. (1978) *Gyn/ecology: The Metaethics of Radical Feminism*, London: The Women's Press.

Dankleman, I. and Davidson, J. (1998) *Women and the Environment in the Third World: Alliance for the Future*, London: Earthscan in association with IUCN.

Dawkins, R. (1976) *The Selfish Gene*, Oxford: Oxford University Press.

De Groot, G. J. (1999) 'A force for change', *Guardian*, 14 June.

Denny, C. (1998) 'Pay gap between the sexes widens', *Guardian*, 16 October.

Department of the Environment, Transport and the Regions (1998) *Opportunities for Change: Consultation Paper on a Revised Strategy for Sustainable Development*, London: HMSO.

Dobson, A. (1999) *Fairness and Futurity: Essays on Environmental Sustainability*, Oxford: Oxford University Press.

Doyle, T. and McEachern, D. (1998) *Environment and Politics*, London: Routledge.

Dreze, J. and Sen, A. (1989) *Hunger and Public Action*, Oxford: Oxford University Press.

Dumayne-Peaty, L. and Wellens, J. (1998) Gender and Physical Geography in the United Kingdom, *Area* 30, 3.

Easlea, B. (1986) 'The masculine image of science with special reference to physics: how much does gender really matter', in Harding, J. (ed.) *Perspectives on Gender and Science*, Brighton: The Falmer Press.

England, K. (1994) 'Getting personal: reflexivity, positionality, and feminist research', *Professional Geographer* 46, 1: 80–89.

European Commission, EOPPSU (1994) *Gender Issues in the Decision Making Process with regard to Urban Space and Housing in Schools of Architecture and at Institutional Level*, Brussels: EC.

European Companion (1998) London: DPR Publishing.

Fagnani, J. (1993) 'Life course and space: dual careers and residential mobility among upper-middle-class families in the Isle-de-France region', in Katz, C. and Monk, J. (eds) *Full Circles, Geographies of Women over the Life Course*, London: Routledge.

Federation of Canadian Municipalities, International Office (1997) *A City Tailored to Women: the Role of Municipal Governments in Achieving Gender Equality*, Montreal: FCM.

Foreman, A. (1999) *Georgiana, Duchess of Devonshire*, London: Flamingo.

Gabizon, S. (1998) 'A dying sea and a dying people', in *Women, Health and Environment*, Utrecht, The Netherlands: Women in Europe for a Common Future.

Garcia-Guadillar, M.-P. (1993) 'Ecologia: women, environment and politics in Venezuela', in Radcliffe, S. and Westwood, S. (eds) *Viva: Women and Popular Protest in Latin America*, London: Routledge.

Garcia-Ramon, M. D. and Caballe, A. (1996) 'From family farming to rural tourism? Gender roles, household survival strategies and environmental perception in rural Spain', First European Urban and Regional Studies Conference, Exeter, April.

Garcia-Ramon, M.D. and Monk, J. (1996) *Women of the European Union: The Politics of Work and Daily Life*, London: Routledge.

Garrett, L. (1995) *The Coming Plague*, Harmondsworth: Penguin Books.

Ghazi, P. and Jones, J. (1996) *Getting a Life*, London: Hodder & Stoughton.

Gibbs, L. (1998) *Love Canal: The Story Continues*, Gabriola Island, BC, Canada: New Society Press.

Giddens, A. (1994) *Beyond Left and Right: The Future of Radical Politics*, Cambridge: Polity Press.

Greed, C. (1994) *Women and Planning: Creating Gendered Realities*, London: Routledge.

Greenfield, S. (1997) 'Unbelievable', *Independent on Sunday*, 6 July.

Greer, G. (1991) *The Change: Women, Ageing and the Menopause*, London: Hamish Hamilton.

Gregson, N. and Lowe, M. (1994) *Servicing the Middle Classes: Class, Gender and Waged Domestic Labour in Contemporary Britain*, London: Routledge.

Gregson, N., Kothari, U., Cream, J., Dwyer, C., Holloway, S., Maddrell, A. and Rose, G. (1997) 'Gender in feminist geography', in Women and Geography Study Group *Feminist Geographies: Explorations in Diversity and Difference*, Harlow: Longman.

Guha, R. and Martinez-Alier, J. (1997) *Varieties of Environmentalism: Essays North and South*, London: Earthscan.

Hall, N. (1997) 'Women want to know the secrets of the universe, too', *Independent on Sunday*, 16 March.

Hallen, P. (1994) 'Re-awakening the exotic: why the conservation movement needs eco-feminism', *Habitat Australia: The Magazine of the Australian Conservation Foundation* 22, 1.

Hanson, S. and Pratt, G. (1995) *Gender, Work and Space*, London: Routledge.

Haraway, D. (1978) 'Animal sociology and a natural economy of the body politic, Part II: the past is the contested zone', in Keller, E.F. and Longino, H.E. (eds) (1996) *Feminism and Science*, Oxford: Oxford University Press.

Haraway, D. (1988) 'Situated knowledges: the science question in feminism and the privilege of partial perspective', in Keller, E.F. and Longino, H.E. (eds) (1996) *Feminism and Science*, Oxford: Oxford University Press.

Harding, S. (1986) *The Science Question in Feminism*, New York: Cornell University Press, and Milton Keynes: Open University Press.

Harding, S. (1990) 'Feminism and theories of scientific knowledge', in Evans, M. (1994) (ed.) *The Woman Question*, London: Sage.

Harries, J.E., Sheahan, D.A., Jobling, S., Mattheissen, P., Neall, P., Sumpter, J., Tylor, T. and Zaman, N. (1997) 'Estrogenic activity in five United Kingdom rivers detected by measurement of vitellogenesis in caged male trout', *Environmental Toxicology and Chemistry* 16, 3: 534–542.

Hartsock, N. (1987) 'The feminist standpoint: developing the ground for a specifically feminist historical materialism', in Harding, S. (ed.) *Feminism and Methodology*, Bloomington, IN: Indiana University Press, and Milton Keynes: Open University Press.

Hashemi, S.M. (eds) and Morshed, L. (1997) 'Grameen Bank: a case study', in Wood, G.D. and Sharif, I.A. (eds) *Who Needs Credit? Poverty and Finance in Bangladesh*, London: Zed Books.

Hayford, A. (1974) 'The geography of women: an historical introduction', *Antipode* 6, 2.

HMSO (1994) *Sustainable Development: The UK Strategy*, London: HMSO.

HMSO (1996) *This Common Inheritance: UK Annual Report*, London: HMSO.

HMSO (1996) *Social Trends*, London: HMSO.

HMSO (1998) *Social Trends*, London: HMSO.

Horelli, L. (1995) 'Self-planned housing and the reproduction of gender and identity', in Ottes, L., Poventud, E., van Schendelen, M. and Segond von Banchet, G. (eds) *Gender and the Built Environment*, Assen, The Netherlands: Van Gorcum.

Horelli, L. and Vepsa, K. (1994) 'In search of supportive structures for everyday life', in Altman, I. and Churchman, A. (eds) *Women and the Environment, Human Behaviour and Environment: Advances in Theory and Research*, vol. 13, London: Plenum Press.

Humbeck, E. (1996) 'The politics of cultural identity: Thai women in Germany', in Garcia-Ramon, M. D. and Monk, J. (eds).

Jackson, C. (1993) 'Women/nature or gender/history? A critique of eco-feminist "development"', *Journal of Peasant Studies* 20, 3: 389–419.

Jacobson, J. (1993) 'Women's work: why development isn't always good news for the second sex', *Foreign Service Journal* (January): 26–31.

James, S. (1995) *The Global Kitchen, The Case for Governments Measuring and Valuing Unwaged Work*, London: Crossroads Books.

Joekes, S., Green, C. and Leach, M. (1996) *Integrating Gender into Environmental Research*, Brighton: IDS Publications.

Katz, C. (1993) Growing girls/closing circles: limits on the spaces of knowledge in rural Sudan and US cities', in Katz, C. and Monk, J. (eds) *Full Circles: Geographies of Women over the Life Course* London: Routledge.

Keller, E.F. (1982) 'Feminism and science', in Keller, E.F. and Longino, H.E. (eds) (1996) *Feminism and Science*, Oxford: Oxford University Press.

Keller, E.F. (1983) *A Feeling for the Organism: The Life and Work of Barbara McClintock*, New York: W.H. Freeman.

Keller, E.F. (1986) 'How gender matters, or why it's so hard for us to count past two', in Harding, J. (ed.) *Perspectives on Gender and Science*, Brighton: The Falmer Press.

Keller, E.F. (1991) 'Language and ideology in evolutionary theory: reading cultural norms into natural law', in Keller, E.F. and Longino, H.E. (eds) (1996) *Feminism and Science*, Oxford: Oxford University Press.

Keller, E. F. and Longino, H. E. (eds) (1996) *Feminism and Science* Oxford: Oxford University Press.

Kempson, F., Bryson, A. and Rowlingson, K. (1994) *Hard Times*, London: Policy Studies Institute.

Koncz, K. (1994) 'Women in the Process of the Political and Economic Changeover', Unpublished paper, Budapest, Hungary: Budapest University of Economic Sciences.

Kranendonk, M. (1997) Pollution and Health in the Ukraine, *Women, Health and Environment*, Newsletter 1–2 (Autumn).

Lane, E., McKay, D. and Newton, K. (1997) *Political Data Handbook: OECD Countries*, 2 edn Oxford: Oxford University Press.

Leyson, B. (1996) 'Medicalisation of menopause: from "feminine forever" to "healthy forever"', in Lykke, N. and Braidotti, R (eds) *Between Monsters, Goddesses and Cyborgs – Feminist Confrontations with Science, Medicine and Cyberspace*, London: Zed Books.

LGMB (1998) *Survey of Local Authority Councillors 1997*, London: LGMB.

Lister, R. (1996) 'Citizenship engendered', in Taylor, D. (ed.) *Critical Social Policy*, London: Sage.

Lloyd, G. (1993) 'Reason, science and the domination of matter', in Keller, E.F. and Longino, H.E. (eds) (1996) *Feminism and Science*, Oxford: Oxford University Press.

Lovenduski, J. and Randall, V. (1993) *Contemporary Feminist Politics*, Oxford: Oxford University Press.

Maguire, S. (1998) 'Gender differences in attitudes to undergraduate fieldwork', *Area* 30, 3.

Martin, E. (1991) 'The egg and the sperm: how science has constructed a romance based on stereotypical male–female roles', in Keller, E.F. and Longino, H.E. (eds) (1996) *Feminism and Science* Oxford: Oxford University Press.

McDowell, L. (1986) 'Beyond patriarchy: a class-based explanation of women's subordination', *Antipode* 18, 3.

McEwan, C. (1997) '"Unfeminine exploits": gender, science and geography in the late nineteenth century', conference paper given at the Royal Geographic Society with the Institute of British Geographers' Annual Conference, Exeter, UK.

McSweeny, B.G. and Freedman, M. (1982) 'Lack of time as an obstacle to women's education: the case of Upper Volta', in Kelly, G. and Elliot, C. (eds) *Women's Education in the Third World: Comparative Perspectives*, Albany, NY: State University of New York Press.

Mellor, M. (1996) 'Sustainability: a feminist approach', in Buckingham-Hatfield, S. and Evans, B. (eds) *Environmental Planning and Sustainability*, Chichester: John Wiley.

Merchant, C. (1983) *The Death of Nature: Women, Ecology and the Scientific Revolution*, New York: Harper & Row.

Merchant, C. (1992) *Radical Ecology: The Search for a Livable World*, London: Routledge.

Merchant, C. (1996) *Earthcare: women and the environment*, London: Routledge.

Mies, M. and Shiva, V. (1993) *Ecofeminism*, London: Zed Books.

Monk, J. and García-Ramon, M.D. (1996) 'Placing women in the European Union', in García-Ramon, M.D. and Monk, J. *ibid.*

Nield, T. (1998) 'Women in the geological society', *Geoscientist* 8, 10.

Office for National Statistics (1997) *Labour Force Survey Quarterly Bulletin* 20, June, London: HMSO.

Osborn, D. and Bigg, T. (1998) *Earth Summit II: Outcomes and Analysis*, London: Earthscan.

Oudshoorn, N. (1996) 'The decline of the one-size-fits-all paradigm, or how reproductive scientists try to cope with post-modernity', in Lykke, N. and Braidotti, R. (eds) *Between Monsters, Goddesses and Cyborgs: Confrontations with Science, Medicine and Cyberspace*, London: Zed Books.

Pepper, D. (1996) *Modern Environmentalism: An Introduction*, London: Routledge.

Pettman, J. (1996) *Worlding Women: A Feminist International Politics*, London: Routledge.

Pietila, H. and Vickers, J. (1996) *Making Women Matter: The Role of the United Nations*, London: Zed Books.

Plumwood, V. (1993) *Feminism and the Mastery of Nature*, London: Routledge.

Poklewski Koziell, S. (1999) 'Two women of the soil', *Resurgence* 195.

Pratt, G. and Hanson, S. (1991) 'On the links between home and work: family–household strategies in a buoyant labour market', *International Journal of Urban and Regional Research* 15, 1.

Pratt, G. and Hanson, S. (1993) 'Women and work across the life course: moving beyond essentialism', in Katz, C. and Monk, J. (eds) *Full Circles: Geographies of Women over the Life Course* London: Routledge.

PRISM (Policy Research in Science) (1995) *Women in Science*, London: The Wellcome Trust.

Pulido, L. (1996) 'A critical review of the methodology of environmental racism research', *Antipode* 28, 2: 142–159.

Pulsipher, L. (1993) '"He won't let she stretch she foot": gender relations in West Indian houseyards', in Katz, C. and Monk, J. (eds) *Full Circles: Geographies of Women over the Life Course*, London: Routledge.

Rahman, R.I. (1997) 'Poverty, profitability of micro enterprises and the role of credit', in Wood, G.D. and Sharif, I.A. (eds) *Who Needs Credit? Poverty and Finance in Bangladesh*, London: Zed Books.

Randall, V. (1987) *Women and Politics: An International Perspective*, 2nd edn Basingstoke: Macmillan.

Rayner, J. (1998) 'The third sex; when the baby is not a boy. Nor a girl', *Observer Life* 01.03.98.

Robinson, D. (1998) 'Differences in Occupational Earnings by Sex', *International Labour Review* 137, 1.

Rose, G. (1993) *Feminism and Geography: The Limits of Geographical Knowledge*, Cambridge: Polity Press.

Sabate-Martinez, A. (1996) 'Rural industrialisation in Spain', in Monk, J. and Garcia-Ramon, M.D.

Said, E. (1995) *Orientalism*, Harmondsworth: Penguin Books.

Salleh, A. (1995) 'Nature, women, labour, capital: living the deepest contradiction', *Capital, Nature, Society* 6, 1.

Seager, J. (1993) *Earth Follies: Feminism, Politics and the Environment*, London: Earthscan.

Shiva, V. (1993) 'The Chipko women's concept of freedom', in Mies, M. and Shiva, V. (eds).

Shrivastava, P. (1992) *Bhopal, Anatomy of a Crisis*, 2nd edn, London: Paul Chapman Publishing.

Simon, J. (1994) 'Opening statement', in Myers, N. and Simon, J. *Scarcity or Abundance – a Debate on the Environment*, New York/London: W.W. Norton.

Simmons, I. (1997) *Humanity and Environment*, Harlow: Addison, Wesley, Longman.

Smith, D. (1974) 'Women's perspective as a radical critique of sociology', in Keller, E.F. and Longino, H.E. (eds) (1996) *Feminism and Science*, Oxford: Oxford University Press.

Stichter , S. (1988) 'Women, employment and the family', in McDowell, L. and Sharp, J. P. (eds) (1997) *Space, Gender, Knowledge: Feminist Readings*, London: Arnold.

Sumpter, J. and Jobling, S. (1995) 'Vittellogenesis as a biomarker for estrogenic contamination of the aquatic environment', *Environmental Health Perspectives* 103 (Suppl 7): 173–178.

Tiles, M. (1987) 'A science or Mars or Venus?', in Keller, E.F. and Longino, H.E. (eds) (1996) *Feminism and Science*, Oxford: Oxford University Press.

Townsend, H. (1993) 'Gender and the life course on the frontiers of settlement in Colombia', in Katz, C. and Monk, J. (eds) *Full Circles: Geographies of Women over the Life Course*, London: Routledge.

United Nations Commission on Environment and Development (1992) *Agenda 21*, Geneva: UNCED.

United States Bureau of the Census (1995) *Statistical Abstract of the USA*, Washington DC: USBC.

Ussher, J. (1997) 'Women on the edge of a breakthrough' *Times Higher*, 4 April.

Vepsa, K. and Horelli, L. (1995) 'Women's networks as a strategy to integrate physical and social aspects in planning', in Ottes, L., Poventud, E., van Schendelen, M. and Segond von Banchet, G. (eds) *Gender and the Built Environment*, Assen, The Netherlands: Van Gorcum.

Walby, S. (1986) *Patriarchy at Work: Patriarchal and Capitalist Relations in Employment*, Cambridge: Polity Press.

Waring, M. (1988) *Counting for Nothing: What Men Value and What Women Are Worth*, Wellington, New Zealand: Allen & Unwin.

Waring, M. (1995) *Who's Counting?*, National Film Board of Canada, video.

Warren, K. (1994) *Ecological Feminism*, London: Routledge.

Wertheim, M. (1997) *Pythagoras' Trousers: God, Physics and the Gender Wars*, London: Fourth Estate.

Wickramasinghe, A. (1995/1996) 'State intervention, women in development, and social changes in Sri Lanka', *Regional Development Studies* 2, Winter.

Wickramasinghe, A. (1997) *Land and Forestry: Women's Local Resource-based Occupations for Sustainable Survival in South Asia*, Peradinaya, Sri Lanka: CORRENSA.

Wickramasinghe, A. (1999) 'Equal opportunities and local initiatives in Sri Lanka', in Buckingham-Hatfield, S. and Percy, S. (eds) (1999).

Woolf, V. (1992) *Three Guineas*, Oxford: Oxford University Press.

Index